A
NATURALIST'S CALENDAR

A
NATURALIST'S CALENDAR

kept at

Swaffham Bulbeck, Cambridgeshire

by

LEONARD BLOMEFIELD

(formerly Jenyns)

SECOND EDITION

Edited by

Sir FRANCIS DARWIN

Honorary Fellow of Christ's College, Cambridge

Cambridge

at the University Press.

1922

CAMBRIDGE
UNIVERSITY PRESS

University Printing House, Cambridge CB2 8BS, United Kingdom

Cambridge University Press is part of the University of Cambridge.

It furthers the University's mission by disseminating knowledge in the pursuit of education, learning and research at the highest international levels of excellence.

www.cambridge.org
Information on this title: www.cambridge.org/9781316619872

First edition 1903
Second edition 1922
First paperback edition 2016

A catalogue record for this publication is available from the British Library

ISBN 978-1-316-61987-2 Paperback

INTRODUCTION.

WHEN Mr Blomefield was preparing his *Observations in Natural History*, of which the Calendar forms a part, he wrote to Mr Darwin asking for contributions. Writing in reply[1] on Oct. 12, 1845, Mr Darwin, after expressing his regret that he cannot add a single fact to it, goes on:—"My work on the species question has impressed me very forcibly with the importance of all such works as your intended one, containing what people are pleased generally to call trifling facts. These are the facts which make one understand the working or economy of nature." These words refer rather to the notes on the habits of birds and beasts than to the Calendar, which is but a skeleton record, and bears the same relation to a treatise on Natural History that Bradshaw's Railway Guide bears to a book of travels. But a Calendar if wisely used is not a dull production; as Darwin says[2] "I think it is very amusing to have a list before one's eyes of the order of appearance of the plants and animals around one; it gives a fresh interest to each fine day." It was

[1] *Life and Letters* ii., pp. 31, 32.

[2] *Life and Letters* i., p. 353. Darwin here refers to the *Naturalist's Pocket Almanack* edited by Mr Jenyns for Van Voorst, which appeared for 1843 to 1847.

to a great extent this point of view that induced me to suggest the republication of Mr Blomefield's Calendar.

I find many of my friends—and not necessarily those whose work lies in science—are interested in knowing when to expect the nightingale, or the cuckoo, or when to look for the pasque-flower on Balsham Dyke. And when we have a record made within a few miles of Cambridge, by one of the most accurate of men, the Syndics of the Press are surely right in making it once more easily accessible[1].

[1] Other Calendars have been compiled. Benjamin Stillingfleet has published at the end of his *Calendar of Flora* 1761 a Calendar "extracted chiefly from Theophrastus' History of plants and put together in the best manner I was able from imperfect materials." Theophrastus gives the appearance of the nightingale as between March 11 and 26, or about a month earlier than in England.

In the same volume of Stillingfleet's is given the Calendar of Alexander Berger made at Upsala in 1755, and published in the *Amœnitates Academicæ* of Linnæus. Berger gives some unexpected information such as "Jan. 2. Wooden walls snap in the night." "Mar. 19. Eves drop towards the noontide sun." But from April onwards he gives the dates of a large number of phenomena. Stillingfleet himself compiled a Calendar in Norfolk for the same year, 1755, between January 5 to October 26. The following are also worthy of mention. *Miscellaneous Tracts and Collections relating to Natural History, selected from the principal writers of Antiquity on that Subject* by W. Falconer, M.D. Cambridge, printed by J. Archdeacon and J. Burges, Printers to the University, 1793, 4to, contains calandars for Greece and Italy, compiled chiefly from Theophrastus, and Columella, together with some more modern data on Aleppo and Italy. A *Naturalist's Calendar* was "extracted from the papers of the late Rev. Gilbert White" by Dr John Aikin and published in 1795; the earliest and latest dates of the numerous phenomena observed, are given, but not a mean date.

The Botanist's Calendar and Pocket Flora, 2 vols., London, 1797, is intended among other things to "assist the investigating Fair One." *Calendarium Botanicum or a Botanical Calendar...of all the British*

But besides ministering to the wants of the amateur Naturalist, such lists as Blomefield's have a real use as

Plants...arranged according to their time of flowering, by the Rev. William Phelps, London, 1810, 8vo. *The Pocket Encyclopædia of Natural Phenomena...* compiled from the MSS. of the late T. F. Forster, by T. Forster, London, 1827, contains a "Rustic Calendar" in which a Saint's name is given for each day, together with various natural phenomena and fragments of folk-lore. I have not been able to see L. Jenyns' *Naturalist's Pocket Almanack*: published by Van Voorst from 1843 to 1847.

Quetelet's *Observations des Phénomènes Périodiques*, Nouveaux Mémoires Acad. R. Bruxelles, T. xv. 1842, contains a mass of observations on times of flowering. The *Observations* are continued in later volumes of the Mémoires.

Normaler Blüthen Kalender von Oesterreich reducirt auf Wien by K. Fritsch, in the Denkschriften of the Vienna Academy, vol. XXVII., 1867, and subsequent years. There are also numerous other phœnological papers by Fritsch in the publications of the Academy. *On a method of registering Natural History observations* by Alfred Newton, Transactions of the Norfolk and Norwich Naturalists' Society, 1870—71, p. 24, contains an interesting plate giving in *facsimile* the authors' shorthand method of recording the phenomena of bird-life: my attention was called to some of the above books by Mr B. D. Jackson of the Linnean Society to whose kindness I am much indebted. The titles of other calendars are given in his *Guide to the Literature of Botany*, London (Index Society) 1881, 8vo, pp. 213, 501.

Lastly may be mentioned the unpublished notes of Thomas Gray, the Poet, preserved in the Library of Pembroke College. I am indebted to Mr Whibley for calling my attention to these, and to Mr Minns for the opportunity of examining them. Gray's notes are in two pocket diaries, viz. *The Gentleman's and Tradesman's Pocket Assistant or Daily Remembrancer for the year* MDCCLV. *A new and Compleat Memorandum Book for General use of all Persons*, and a similar, but not identical, book for 1760. The pocket books contain entries of money received and expended, records in Latin of the state of his health, as well as the natural history notes, with which alone we are concerned. He seems to have been especially interested in the weather, and usually gives the direction of the wind and the general character of the day. His observations are irregular and do not

contributions to the science of Phœnology. This is a science which has suffered in two ways—viz. from incompleteness in observation, and a too bold style of theorising. In illustration of the first point we may mention Askenasy's observations[1] on the growth of wild-cherry buds. An ordinary phœnologist might have been content with noting the date of flowering of the cherry. But Askenasy took the weight of 100 buds at various times through the year and thus got a true idea of the development of the flower. He also tested the reaction of the buds to warmth at various seasons, showing that in the early autumn they cannot be forced into bloom, whereas in early spring they can be so forced. In this way he showed that there is an inherent periodicity in the plant which cannot be accounted for simply by the direct action of external conditions. It is not meant that the natural changes of temperature through the year have no effect on the periodic phenomena, this is amply contradicted in every garden, merely

furnish anything like a complete Calendar of Nature. In the 1755 book he is diligent up to the end of April, some record occurring practically every day, while on a fair number of days there are numerous entries. The following may serve as examples:

Mar. 2. Chaffinches singing.

Mar. 5. Flies creep abroad and moths fly in the evening. Rooks building.

Mar. 10. Bazelman-Narcissus blows, 9 inches high (3 flowers) and the leaves 25 inches long.

Ap. 3. Lilac, sweetbryar, and dwarf Almond put out leaves. Snowdrops gone off, and crocus going.

For an interesting series of the Poet's phœnological notes, see *The Works of Gray* edited by Edmund Gosse, 1884, vol. iii, pp. 88—96.

[1] *Botanische Zeitung* 1877.

that we cannot account for the phenomena entirely by the course of the temperature. It was in this direction that the earlier phœnologists went wrong. They combined time and temperature by, for instance, multiplying the number of days between the sowing of barley and harvest time by the average temperature of that period. Thus in Bavaria the average temperature was 17·25, the time was 100 days, giving a product of 1725. In Cumbal the temperature was 10·7, the period 168 days, making 1798: and similar constants for other regions agree fairly. Formulæ were also constructed in which the square of the temperature or the square of the time was made use of. These calculations may perhaps allow correct predictions to be made in some cases, but as parts of a wide treatment of periodic phenomena they have little or no value. Krašan[1] has given some interesting examples of plants whose periods do not fall within any known rule. Thus *Spiranthes autumnalis* flowers later in southern than in northern regions: *Artemisia* and *Calluna* flower at the same time in the north and south. Others flower earlier in the south, but not in proportion to the amount of heat to which they have been exposed. Others again flower earlier in southern regions although the sum of heat received is not greater there than in somewhat more northern places. These instances are chiefly of interest to us as showing that the subject is a difficult and complex one, and that to obtain generalisations of scientific value is no easy matter. But such considerations need in no way detract from the interest of Mr Blomefield's work, his careful observations have a permanent value.

[1] Engler's *Bot. Jahrb.* III.

His introductory remarks, "On the importance of registers of periodic phenomena" show how much there is to interest the general reader. He points out[1] that "however much the seasons may differ in different years, the phenomena generally follow one another in the same order. This indeed is what we might expect, from the circumstance of any interruption in the time of their occurrence, due to seasonal influence, necessarily affecting them all equally. And it follows that those which occur together any one year, will occur at or nearly [at] the same time every other."

He gives examples of this: thus, on an average, the box-tree and the ground-ivy (*Nepeta glechoma*) open their flowers together on April 3[2]. While in 1845, a backward year, they flower later, but still close together, viz. April 20 and April 19. Or to extend the coincidences to the animal kingdom, the black-cap does not appear in the north of England till the larches are visibly green, and the wood-warbler make its appearance with the leafing of the oaks and elms.

The Calendar may be respectfully applied to test the accuracy of poets. Thus Shakspeare is quite right in making the daffodil come before the swallow dares, since the latest daffodil flowers on April 4, and the swallow does not appear till April 9 at the earliest. Browning[3] is safe too, in letting the chaffinch sing "on the orchard bough" now "That the lowest boughs and the brush-wood sheaf Round the elm-tree bole are in tiny leaf." For the chaffinch begins his song in

[1] *Observations in Natural History*, p. 332.

[2] Further observations give March 28 and 30 for these flowers.

[3] "Home-thoughts from abroad."

February and the elm's leaves show on April 9. The "blossomed pear tree in the hedge" a line or two later points to April 13. Our Calendar may be used too for a mild antiquarianism. Thus Blomefield[1] says that the "*flos-cuculi* or cuckoo flower of the older botanists was so called from its opening its flowers about the time of the cookoo's commencing his call." The question is what flower is meant: the older botanist referred to is probably Gerarde and he seems to mean *Cardamine pratensis* known as Lady's Smock or the Cuckoo Flower. Now the cuckoo begins his song April 29, and *Cardamine* blossoms April 19. The coincidence is therefore only moderately close, though in Gilbert White's Calendar they come together thus:—*Cardamine* April 6—20, Cuckoo April 7—26.

Wood-sorrel, *Oxalis acetosella,* may have been intended with about equal propriety. Wood-sorrel was known in mediæval latin as *panis cuculi* and as Cuckoo-sorrel[2] by the Saxons, and in B. Stillingfleet's *Calendar of Flora* (1755) it is said to flower on April 16, while the cuckoo begins his notes on April 17. It is interesting to find that in the Swedish Calendar the cuckoo sings on May 12 and the wood-sorrel flowers on May 13. The question remains, could it have been the Cuckoo-pint (*Arum maculatum*) usually known as Lords and Ladies? This is given as flowering at the right time, May 1—but according to Prior[3] the derivation of the name is of a different kind. *Lychnis flos-cuculi* the Ragged Robin flowers May 19, and may be left out of consideration.

[1] *Observations in Natural History*, p. 334.

[2] Prior's *Popular Names of British Plants*, Ed. III. 1879.

[3] *Ibid.*

Blomefield suggests another possible use for a natural history calendar, namely the regulation of the operations of husbandry by the sequence of nature rather than by dates. "Thus," he writes[1], "the middle of March may be, in the long run, the most suitable time for sowing several kinds of grain." And after pointing out how easily the husbandman may go wrong if he trusts to this date, he continues, "But if he know that the same conditions of soil and atmosphere, which are requisite for his purpose, are also requisite for bringing into flower or leaf any particular plant, he cannot be far out in his reckoning, if he wait for the first appearance of such plant to guide him in his operations." He thus gives his sanction to the ancient Greek custom. Stillingfleet quotes from the *Birds* of Aristophanes that "the crane points out the time of sowing" and the appearance of the kite tells you "when it is time to shear your sheep." And for a more modern example we have, on Solander's authority[2], a Swedish proverb "when you see the white wagtail you may turn your sheep into the fields; and when you see the wheat-ear you may sow your grain" The poet Gray[3] used this method when he told his friends not to expect him in Cambridge "till the codlin hedge at Pembroke was out in blossom."

As regards the author of the Calendar we fortunately have, in his privately printed autobiographical notes[4], the material for a few words on his life.

[1] *Observations in Natural History*, p. 335.

[2] Stillingfleet, pp. 3, 5. The Greek folk lore is from p. viii.

[3] *English Men of Letters, Gray*, by Edmund Gosse, 1889, p. 154.

[4] *Chapters in my Life.* With appendix, containing special notices of particular incidents and persons: Leonard Blomefield (late Jenyns).

Leonard Jenyns was born "at 10 p.m." on May 25, 1800, "in a house in Pall Mall, London, now pulled down, but then occupied by the celebrated Dr [William] Heberden," who was his uncle on the mother's side. His father was the Rev. George Leonard Jenyns[1] of Bottisham Hall which he inherited from Soame Jenyns, known as the author of *A free inquiry into the nature and origin of evil*, London, 1757. *A view of the internal evidence of the Christian Religion*, London, 1776. Leonard[2] the younger went to a private school at Putney and in 1813 to Eton. He writes (p 6) of his boyhood, "I also, as a boy had that fondness for order, method and precision, which I retained through life; arranging all my things, clothes, books, etc. with great particularity;—neat and tidy in everything. I was likewise somewhat taciturn. My school fellows nicknamed me *Methodist* and Dummy. This I did not like. But it was true all the same. Through life I have been a man of few words, as regards the staple conversation of ordinary society; and even to old age I have been often called a *very particular gentleman*."

For Private Circulation. Bath, 1887. Also an undated pamphlet headed *Addenda*. See also Mr H. H. Winwood's "Reminiscences of the late President and Founder of the Club" (The Rev. Leonard Blomefield), read before the Bath Natural History and Antiquarian Field Club, January 10, 1894.

[1] Mr Jenyns the elder lived the life of a country gentleman rather than that of a clergyman, although he was a Prebendary of Ely. He came of what is described (*Works of Soame Jenyns*, 1790, vol. i. p. xv) as "the ancient and respectable family" of Jenyns of Churchill in Somerset.

[2] His change of name occurred in 1871 when the property and surname of Francis Blomefield, "the celebrated historian of Norfolk," devolved upon him.

At Eton he was fond of experimenting and he was on one occasion introduced to Sir Joseph Banks as the "Eton boy who lit his rooms with gas" (p. 33). Mr Blomefield gave me in conversation another link with the past,—that he occasionally drove to Eton in the carriage of Dr Heberden on his way to see the King at Windsor. From Eton he went to St John's College, Cambridge, and at the University he made the acquaintance of Prof. Henslow, with whom he was much associated in natural history pursuits; the two friends were further united by the marriage of Henslow with Miss Harriet Jenyns in 1823. Mr Blomefield afterwards wrote the life of his brother in law[1].

While at Cambridge Blomefield had the chance of going with FitzRoy in the *Beagle*, an offer which he declined after a day of hesitation. He seems to have been to some extent instrumental in choosing Charles Darwin for the post.

At Cambridge he saw a good deal of Darwin who often visited him at his Vicarage of Swaffham Bulbeck, and went with him beetle-catching in the Fens. Darwin says of him in an unpublished passage in his Autobiography, "At first I disliked him from his somewhat grim and sarcastic expression; and it is not often that a first impression is lost; but I was completely mistaken and found him very kind hearted, pleasant and with a good stock of humour."

Mr Blomefield described the fish in the *Zoology of the Voyage of H.M.S. "Beagle"*—and continued to correspond with Darwin[2].

[1] *Memoir of the Rev. John Stevens Henslow*, London, 1862.

[2] See *Life and Letters of Charles Darwin.*

He held the living of Swaffham Bulbeck for nearly thirty years, when he resigned it owing to the ill health of his wife[1]. They moved for a short time to the Isle of Wight and then to Bath, where he remained to the end of his life. There I had the pleasure of being his guest for the meeting of the British Association in 1888 and I remember well his courteous and somewhat formal manner, and the general vigour of his personality—very striking in so old a man. He has left two small indications of his personal appearance which connect him with the past in a curious way. He was as a young man taken by a stranger in a coach for a son of Dr Heberden, and the resemblance was often noticed by his own family[2]. The other instance occurred in later life: Philip Duncan, of New College, Oxford, who died in 1865 at the age of 93, had been in the company of Gilbert White who died in 1793. Blomefield[3] says, "On asking him what sort of man White was—as to height, figure and general appearance—he answered to my great amusement 'O, much such as you are!'" White was one of Blomefield's heroes, at Eton he copied out nearly the whole of the Natural History of Selborne "under the apprehension that I might never see the book again" (p. 43), and in 1843 he brought out an edition of the book with notes. Among Blomefield's other works were a *Manual of British Vertebrate Animals* (Cambridge) 1836: *Observations in Natural History*, 1846, from which the present Calendar is taken: *Observations in Meteorology*, 1858, and a long

[1] Jane Daubeny, a niece of the Oxford professor of Botany.

[2] *Chapters etc.*, pp. 13, 14.

[3] *Addenda*, p. 67.

series of papers in various scientific journals. His zeal for science was also shown by his founding the "Bath Natural History and Antiquarian Field Club." He died at Bath September 1, 1893.

As a Naturalist he was known for his minute and scrupulous exactness in matters of fact. Darwin has said, "Accuracy is the soul of Natural History...absolute accuracy is the hardest merit to attain, and the highest merit." And this quality of Mr Blomefield's work gives a special value to the Calendar here reprinted.

To those who may be inclined to continue Mr Blomefield's work I would venture to suggest the observation of those animals and plants of which he only obtained one or two records. It should be a pleasant task to Cambridge Naturalists to complete the faithful work of such a predecessor.

I desire to express my thanks to the authorities of the Museum of Zoology at Cambridge for allowing me to make use of Mr Blomefield's annotated copy of his Calendar. And I have to thank Mr Rutherford for the assistance he has given me in recalculating the mean dates.

FRANCIS DARWIN.

10 MADINGLEY ROAD,
CAMBRIDGE.
October, 1921.

NOTE ON THE CALENDAR.

Mr Blomefield's Calendar was founded on observations made near Cambridge between the years 1820 and 1831, from these data he calculated the mean date, and recorded the earliest and latest occurrence of each phenomenon. When an observation was made but once the date is "entered under the head of mean, as a place which it is entitled to hold till other observations of the same phenomenon[1]" have been made. "When but two observations have been made, and they have occurred on two consecutive days, these two days have been bracketed together under the head of mean[2]." Mr Blomefield also gave in a separate column his observations for 1845, with a view to showing the actual sequence of phenomena for a single year. In his annotated copy he gave a large number of additional observations for 1846—49. I have thought it best to omit the record of 1845, as a separate entry, and to include its contents in recalculating the means and restating the earliest and latest appearances, with the help of the MS. of 1846—49. In this way it is possible to give, in a good many instances, a mean date founded on the observations of 17 years. The number of years on which each mean is calculated is given in the second column, except in the cases where only a single obser-

[1] *Observations in Natural History*, p. 362.

[2] *Ibid.*

vation was made. The MS. of 1846—49 gives some observations on species which do not occur in the original Calendar, these are included in the present edition.

The index has been enlarged by the addition of the scientific names, because it was found in practice that Mr Blomefield's index, in which "only the English names are given in general" was not satisfactory. It has been found necessary to alter the form of two or three of Mr Blomefield's footnotes in consequence of the present version being based on data not included in the original Calendar.

It is fair to Mr Blomefield's memory to repeat the following words of caution[1] which form part of his introductory pages on the "Importance of Registers of Periodic Phenomena." "The bulk of the observations having been made and registered many years back, it is extremely probable that there may be some wrong entries by mistake in the journal from which they are now copied, though it is believed that they are generally correct. It will also sometimes happen that the first occurrence of a particular phenomenon is in reality previous to the day of its being first noticed; and this is especially likely to be the case with respect to the appearance of birds and insects, which cannot be watched with the same exactness as plants, and in regard of which the attention is not given to any particular individuals but to the species generally."

[1] *Observations in Natural History*, p. 365.

CALENDAR.

CALENDAR OF PERIODIC PHENOMENA IN NATURAL HISTORY,

AS OBSERVED IN THE NEIGHBOURHOOD OF SWAFFHAM BULBECK.

Abbreviations used:

ap.	first appearance.	*sg. reas.*	song reassumed.
sg. com.	song commences.	*l.*	first opening of leaves.
sg. ceas.	song ceases.	*fl.*	first opening of flowers.

Italic characters are used for the scientific names: where these alone occur, there are no English names generally received.

The time of flowering (*fl.*) is determined by the visibility of the anthers.
The time of leafing (*l.*) by the exposure of the upper surface of the leaves.

JANUARY.

Phenomena.	Date of occurrence: Mean ([a])		Earliest	Latest
Song Thrush (*Turdus musicus*), sg. com.	Jan. 4	17	([b]) Jan. 1	Feb. 26
Wren (*Troglodytes europæus*), sg. com.	Jan. 5	13	Jan. 1	Jan. 18
Redbreast (*Erithaca rubecula*), sg. com.	Jan. 7	13	Jan. 1	Jan. 28
Common Bunting (*Emberiza miliaria*), note com. ...	Jan. 15	7	Jan. 1	Feb. 4
Marsh Titmouse (*Parus palustris*), note com. ...	Jan. 18	12	Jan. 1	Feb. 17
Hedge Sparrow (*Accentor modularis*), sg. com. ...	Jan. 18	16	Jan. 1	Feb. 15

Winter Gnat (*Trichocera hiemalis*), ap.	Jan. 20	6	Jan. 5	Feb. 18
Pale perfoliate Honeysuckle (*Lonicera caprifolium*), l.	Jan. 21	6	Jan. 1	Feb. 22
Skylark (*Alauda arvensis*), sg. com.	Jan. 22	17	Jan. 7	Feb. 21
Mezereon (*Daphne mezereum*), fl.	Jan. 22	2	Jan. 11	Feb. 2
Cole Titmouse (*Parus ater*), note com.	Jan. 23	8	Jan. 7	Feb. 20
Furze (*Ulex europæus*), fl.	Jan. 24	15	Jan. 1	Apr. 4
Great Titmouse (*Parus major*), sg. com.	Jan. 26	16	Jan. 7	Feb. 22
Hazel (*Corylus avellana*), fl. (c)	Jan. 26	16	Jan. 1	Feb. 20
Hepatica (*Hepatica triloba*), fl.	Jan. 26	13	Jan. 1	Mar. 25
Winter Aconite (*Eranthis hiemalis*), fl.	Jan. 26	15	Jan. 1	Feb. 13
Starlings (*Sturnus vulgaris*), resort to buildings ...	Jan. 27	5	Jan. 9	Feb. 15
Stinking Hellebore (*Helleborus fœtidus*), fl.	Jan. 27	17	Jan. 2	Mar. 6
Missel Thrush (*Turdus viscivorus*), sg. com.	Jan. 29	17	Jan. 1	Mar. 10
Daisy (*Bellis perennis*), fl.	Jan. 29	16	Jan. 1	Mar. 12
Snowdrop (*Galanthus nivalis*), fl.	Jan. 30	17	Jan. 18	Feb. 16
Peziza coccinea, ap.	Jan. 31	2	Jan. 2	Mar. 1

(a) The small figures in the second column under the head of Mean, indicate the number of years from which the mean is deduced, when the observation has been made more than once.

(b) All those phenomena which are referred to January 1st as the earliest date, may be considered as occasionally shewing themselves in December of the previous year.

(c) This entry refers to the opening of the male catkins: the female blossoms do not usually shew themselves till a few days later.

FEBRUARY.

Phenomena.	Date of occurrence Mean		Earliest	Latest
Flocks of Greenfinches (*Coccothraustes chloris*), separate	Feb. 2			
Tawny Owl (*Syrnium aluco*), hoots	Feb. 2	5	Jan. 24	Feb. 14
Spurge-Laurel (*Daphne laureola*), fl.	Feb. 2	17	Jan. 3	Mar. 7
Chaffinch (*Fringilla cælebs*), sg. com.	Feb. 4	17	Jan. 7	Feb. 19
Field Speedwell (*Veronica agrestis*), fl.	Feb. 4	7	Jan. 3	Apr. 5
House-flies ap. on windows ([a])...	Feb. 4	9	Jan. 3	Mar. 15
Hive Bee (*Apis mellifica*) comes abroad	Feb. 5	16	Jan. 1	Mar. 31
Pied Wagtail (*Motacilla yarrellii*), first seen... ...	Feb. 8	8	Jan. 2	Mar. 17
Small smooth Eft (*Triton punctatus*), ap. in ponds ...	Feb. 8			
Butcher's-broom (*Ruscus aculeatus*), fl.	Feb. 8	10	Jan. 3	Apr. 1
Gymnostomum ovatum, ripens its capsules	Feb. 9	4	Jan. 2	Mar. 22
House Pigeon (*Columba livia*, var. *domestica*), lays ...	Feb. 11			
Elder (*Sambucus nigra*), l.	Feb. 13	17	Jan. 2	Mar. 22
Japan Quince (*Pyrus japonica*), fl.	Feb. 13	3	Jan. 7	Mar. 20
Primrose (*Primula vulgaris*), fl.	Feb. 13	15	Jan. 3	Mar. 30
Gold-crested Wren (*Regulus cristatus*), sg. com. ...	Feb. 14	14	Jan. 4	Mar. 31
Partridge (*Perdix cinerea*), pairs	Feb. 14	9	Jan. 12	Mar. 12
Double Daisy (*Bellis perennis, fl. pl.*), fl.	Feb. 14	5	Jan. 2	Apr. 13
Yellowhammer (*Emberiza citrinella*), sg. com. ...	Feb. 16	17	Jan. 30	Mar. 4
Jackdaws (*Corvus monedula*), resort to chimneys ...	Feb. 18	5	Jan. 25	Mar. 27

Blackbird (*Turdus merula*), sg. com.	Feb. 19	17	Jan. 18	Mar. 27
Red Dead-nettle (*Lamium purpureum*), fl.	Feb. 19	9	Jan. 2	Mar. 22
Common Honeysuckle (*Lonicera periclymenum*), l. ...	Feb. 20	10	Jan. 23	Mar. 26
Spring Crocus (*Crocus vernus*), fl.	Feb. 20	15	Jan. 31	Mar. 24
Dandelion (*Taraxacum officinale*), fl.	Feb. 21	13	Jan. 1	Apr. 5
Sweet-scented Coltsfoot (*Tussilago fragrans*), fl. ...	Feb. 21	7	Jan. 23	Mar. 29
Velia currens, ap. on the surface of streams	Feb. 22	5	Feb. 9	Feb. 26
Drone Fly (*Eristalis tenax*), ap.	Feb. 23	2	Feb. 8	Mar. 11
Greenfinch (*Coccothraustes chloris*), sg. com.	Feb. 23	17	Feb. 1	Mar. 20
Ring-dove (*Columba palumbus*), coos	Feb. 23	17	Jan. 1	Mar. 26
Earthworms lie out	Feb. 23	8	Jan. 23	Mar. 30
Lesser Periwinkle (*Vinca minor*), fl.	Feb. 24	17	Jan. 2	Apr. 25
Heath Snail (*Helix ericetorum*), comes abroad (b) ...	Feb. 27	6	Jan. 30	Mar. 30
Japan Kerria (*Kerria japonica*), l.	Feb. 27	10	Jan. 31	Mar. 28
Alder (*Alnus glutinosa*), fl.	Feb. 28	12	Jan. 21	Mar. 20
Pilewort (*Ranunculus ficaria*), fl.	Feb. 28	17	Jan. 21	Mar. 28
Yew (*Taxus baccata*), fl.	Feb. 28	17	Feb. 5	Mar. 29

(a) This refers to the period of their being first roused to a state of activity in a room without a fire.

(b) I observe that this species of snail, which is everywhere common, is always much earlier in coming abroad than either the *Helix aspersa* or the *H. nemoralis*.

MARCH.

PHENOMENA.	Date of occurrence: Mean		Earliest	Latest
Flocks of Wild Geese return northwards	Mar. 3	2	Mar. 1	Mar. 5
Ladybird (*Coccinella 7 punctata*), ap.	Mar. 3	12	Jan. 3	May 18
Rooks (*Corvus frugilegus*), build	Mar. 4	17	Feb. 12	Mar. 14
Mistletoe (*Viscum album*), fl.	Mar. 4	8	Feb. 10	Mar. 28
Stock-dove (*Columba œnas*), note com.	Mar. 5	9	Feb. 17	Mar. 27
Cushion-moss (*Grimmia pulvinata*), ripens its capsules	Mar. 5	2	Feb. 13	Mar. 25
Marsh-Marygold (*Caltha palustris*), fl.	Mar. 5	11	Feb. 9	Mar. 29
Persian lilac (*Syringa persica*), l.	Mar. 5	2	Feb. 23	Mar. 16
Sweet Violet (*Viola odorata*), fl.	Mar. 5	17	Jan. 25	Mar. 29
Common Whitlow-grass (*Draba verna*), fl.	Mar. 6	10	Jan. 21	Apr. 10
Turkey-cock (*Meleagris gallopavo*), struts and gobbles ([a])	Mar. 7	12	Feb. 12	Apr. 4
House Pigeon hatches	Mar. 8	2	Feb. 28	Mar. 16
Blue Navelwort (*Omphalodes verna*), fl.	Mar. 8	6	Jan. 8	Apr. 2
Farsetia deltoidea, fl.	Mar. 8	2	Feb. 8	Apr. 5
Hooded Crow (*Corvus cornix*), last seen	Mar. 9	2		
Apricot (*Prunus armeniaca*), fl.	Mar. 9	15	Feb. 19	Mar. 31
Ivy-leaved Speedwell (*Veronica hederæfolia*), fl. ...	Mar. 11	7	Feb. 1	Apr. 3
Daffodil (*Narcissus pseudonarcissus*), fl. ([b])	Mar. 12	17	Feb. 23	Apr. 4
Gooseberry (*Ribes grossularia*), l.	Mar. 12	17	Feb. 24	Mar. 30

Quince (*Cydonia vulgaris*), l.	Mar. 12	2	Feb. 28	Mar. 24
Broods of small Coleopterous Insects on wing ([c]) ...	Mar. 13	6	Feb. 19	Apr. 16
Dor-Beetle (*Geotrupes stercorarius*), ap.	Mar. 13	10	Jan. 30	May 4
Peach (*Amygdalus persica*), fl.	Mar. 13	16	Feb. 20	Mar. 31
White Dead-nettle (*Lamium album*), fl.	Mar. 13	13	Jan. 19	Apr. 22
Gossamer floats ([d])	Mar. 14	5	Feb. 12	Apr. 1
Small bloody-nose Beetle (*Timarcha coriaria*), ap. ...	Mar. 14	5	Jan. 30	June 3
Aspen (*Populus tremula*), fl.	Mar. 14	7	Feb. 14	Mar. 30
Common Coltsfoot (*Tussilago farfara*), fl.	Mar. 14	17	Feb. 23	Apr. 4
Peacock (*Pavo cristatus*), screams ([e])	Mar. 14	5	Feb. 19	Apr. 1
Dog-rose (*Rosa canina*), l.	Mar. 15	8	Feb. 23	Apr. 7
Privet (*Ligustrum vulgare*), l.	Mar. 15	14	Feb. 23	Apr. 12
Snowberry (*Symphoricarpos racemosus*), l.	Mar. 15	3	Mar. 11	Mar. 23
Brimstone Butterfly (*Gonepteryx rhamni*), ap. ...	Mar. 16	17	Feb. 2	May 20
Frog (*Rana temporaria*), spawns	Mar. 16	9	Mar. 4	Mar. 25

([a]) This entry is also made in White's own language;—the phenomenon itself indicates those feelings in the cock bird, which are connected with the approach of the breeding-season.

([b]) In 1834, a remarkably forward season, this plant was in flower on the 28th of January. [The dates are calculated from observations made in 1820—31 and 1845—49.]

([c]) These broods of small coleopterous insects, consisting chiefly of *Curculionidæ* and *Staphylinidæ*, are alluded to in a former part of this work, as coming on wing the first mild spring day that may occur in February or March.

([d]) I have generally observed this phenomenon twice in the year, spring and autumn, the same as White, whose expressive term "floats" is here retained. The autumnal entry will be found further on.

([e]) This does not refer to the ordinary call or cry of the peacock, but to a peculiar scream, uttered only by the male bird, when under the influence of sexual desire, and very characteristic of the first warm weather that occurs in early spring.

MARCH.

Phenomena.	Date of occurrence Mean		Earliest	Latest
Woodcock (*Scolopax rusticola*), last seen	Mar. 16	2	Mar. 10	Mar. 23
Black Currant (*Ribes nigrum*), l.	Mar. 16	6	Feb. 26	Apr. 9
Creeper (*Certhia familiaris*), spring note com. ...	Mar. 16	6	Feb. 3	Apr. 16
Lilac (*Syringa vulgaris*), l.	Mar. 16	17	Feb. 21	Apr. 4
Syringa (*Philadelphus coronarius*), l.	Mar. 16	14	Feb. 23	Apr. 12
Frog (*Rana temporaria*), croaks	Mar. 17	4	Mar. 8	Apr. 2
Humble-Bee (*Bombus*), ap.	Mar. 17	17	Feb. 11	Mar. 31
Pied Wagtail, spring note com.	Mar. 17			
Common Rose (*Rosa centifolia*), l.	Mar. 17	2	Mar. 12	Mar. 22
Oats (*Avena sativa*), sown	Mar. 17	5	Feb. 22	Apr. 21
Common Elm (*Ulmus campestris*), fl.	Mar. 18	17	Feb. 23	Apr. 4
Hygrometric Cord-moss (*Funaria hygrometrica*), ripens its capsules	Mar. 18	4	Jan. 14	Apr. 14
Whirlwig Beetle (*Gyrinus natator*), ap.	Mar. 19	10	Jan. 25	May 11
Willows open their catkins ([a])...	Mar. 19	8	Mar. 11	Mar. 27
Japan Kerria (*Kerria japonica*), fl.	Mar. 20	10	Feb. 24	Apr. 21
Carabus nemoralis, ap.	Mar. 21	5	Feb. 27	Apr. 20
Common Gnat (*Culex pipiens*), ap.	Mar. 21	7	Feb. 21	May 3
Elater lineatus, ap. ([b])	Mar. 21	2	Mar. 7	Apr. 5
Sweet-briar (*Rosa rubiginosa*), l.	Mar. 21	6	Mar. 8	Apr. 3

Badister bipustulatus, ap.	Mar. 22			
Elater sputator, ap.	Mar. 22			
Magpie (*Pica caudata*), builds	Mar. 22			
Weeping Willow (*Salix babylonica*), l.	Mar. 22	11	Feb. 18	Apr. 23
Common Linnet (*Linota cannabina*), sg. com. (c) ...	Mar. 23	17	Feb. 22	Apr. 16
Barley (*Hordeum vulgare*), sown	Mar. 23	4	Mar. 15	Mar. 30
Red Currant (*Ribes rubrum*), l.	Mar. 23	14	Mar. 7	Apr. 12
Small-tortoiseshell Butterfly (*Vanessa urticæ*), ap. ...	Mar. 24	13	Mar. 2	Apr. 15
White Poplar (*Populus alba*), fl.	Mar. 24	8	Mar. 7	Apr. 13
Whitethorn (*Cratægus oxyacantha*), l.	Mar. 24	17	Feb. 27	Apr. 16
Earwig (*Forficula auricularia*), ap.	Mar. 25	3	Feb. 1	Apr. 28
Green Woodpecker (*Picus viridis*), cries	Mar. 25	5	Feb. 25	Apr. 17
Red Ant (*Formica*), ap.	Mar 25	12	Jan. 18	Apr. 24
Rook (*C. frugilegus*), lays	Mar. 25	3	Mar. 17	Apr. 2
Bramble (*Rubus fruticosus*), l.	Mar. 25	9	Feb. 25	Apr. 10
Jackdaw (*Corvus monedula*), builds	Mar. 26	9	Mar. 7	Apr. 7
Dog's Mercury (*Mercurialis perennis*), fl.	Mar. 26	5	Feb. 15	Apr. 10
Common Toad (*Bufo vulgaris*), spawns	Mar. 27	7	Mar. 16	Apr. 5

(a) No particular species is mentioned, as several open their catkins about the same time: the species, too, are with difficulty discriminated from each other. Yet the general phenomenon, as indicative of spring, is worth noticing.

(b) The larvæ of this species, and the *E. sputator* noticed further down, are the *wireworms* so destructive in agriculture, by attacking the roots of plants.

(c) The commencement of song in this species is coincident with the breaking up of the winter flocks, the several individuals then betaking themselves to gardens and shrubberies.

MARCH.

Phenomena.	Date of occurrence: Mean		Earliest	Latest
Harpalus aeneus, ap.	Mar. 27	3	Mar. 12	Apr. 6
Creeping Crowfoot (*Ranunculus repens*), fl. (a) ...	Mar. 27	7	Jan. 6	May 31
Hyacinth (*Hyacinthus orientalis*), fl.	Mar. 27	7	Mar. 7	Apr. 9
Domestic Goose hatches	Mar. 28			
Oil Beetle (*Proscarabæus vulgaris*), ap.	Mar. 28	17	Feb. 25	Apr. 7
Otiorhynchus tenebricosus, ap.	Mar. 28	3	Mar. 13	Apr. 10
Tawny Owl (*Syrnium aluco*), lays	Mar. 28			
Almond (*Amygdalus communis*), fl.	Mar. 28	3	Mar. 20	Apr. 6
Box (*Buxus sempervirens*), fl. (b)	Mar. 28	14	Feb. 22	Apr. 22
Gooseberry (*Ribes grossularia*), fl.	Mar. 29	15	Mar. 12	Apr. 20
Grape-Hyacinth (*Muscari racemosum*), fl.	Mar. 29	6	Mar. 14	Apr. 13
Peach (*Amygdalus persica*), l.	Mar. 29	12	Mar. 7	Apr. 16
Wood-sorrel (*Oxalis acetosella*), fl.	Mar. 29			
Large bloody-nose Beetle (*Timarcha tenebricosa*), ap.	Mar. 30	10	Feb. 19	May 8
Pipistrelle Bat (*Vespertilio pipistrellus*), comes abroad	Mar. 30	6	Mar. 3	May 20
Barberry (*Berberis vulgaris*), l.	Mar. 30	7	Feb. 26	Apr. 19
Ground-ivy (*Nepeta glechoma*), fl.	Mar. 30	16	Feb. 26	Apr. 23
Wych Elm (*Ulmus montana*), fl.	Mar. 30	4	Mar. 23	Apr. 10
Yellow Figwort (*Scrophularia vernalis*), fl.	Mar. 30	6	Mar. 5	Apr. 23
Banded Snail (*Helix nemoralis*), comes abroad ...	Mar. 31	10	Jan. 31	May 17

Domestic Duck (*Anas boschas* var. *domest.*) hatches	Mar. 31	5	Mar. 26	Apr. 15
Apricot (*Prunus armeniaca*), l.	Mar. 31	12	Mar. 10	Apr. 18
Dwarf purple Iris (*Iris pumila*), fl.	Mar. 31	7	Mar. 12	Apr. 23
Hairy Violet (*Viola hirta*), fl.	Mar. 31	9	Mar. 16	Apr. 17
Mealy-tree (*Viburnum lantana*), l.	Mar. 31	14	Mar. 10	Apr. 22
Saxifraga crassifolia, fl.	Mar. 31			
Wall-flower (*Cheiranthus cheiri*), fl.	Mar. 31	8	Jan. 9	Apr. 28

APRIL.

Cowslip (*Primula veris*), fl.	Apr. 1	16	Feb. 5	Apr. 23
Hazel (*Corylus avellana*), l.	Apr. 2	17	Mar. 12	Apr. 22
Horse-chestnut (*Æsculus hippocastanum*), l.	Apr. 2	17	Mar. 15	Apr. 21
Larch (*Larix europæa*), l.	Apr. 2	7	Mar. 21	Apr. 11
Raspberry (*Rubus idæus*), l.	Apr. 2	12	Mar. 5	Apr. 27
Solid-rooted Fumitory (*Corydalis solida*), fl.	Apr. 2	8	Mar. 18	Apr. 17
Spring Bitter-vetch (*Lathyrus vernus*), fl.	Apr. 2			
Burnet Rose (*Rosa spinosissima*), l.	Apr. 3	9	Mar. 4	Apr. 21
Common Laurel (*Prunus laurocerasus*), fl.	Apr. 3	4	Mar. 13	Apr. 26
Ivy (*Hedera helix*), berries ripe	Apr. 3	6	Mar. 14	Apr. 17
House Sparrow (*Passer domesticus*), builds	Apr. 4	7	Mar. 16	Apr. 24

([a]) The period of first flowering in this plant, though a perennial, appears to have a great range, and is very uncertain.
([b]) In 1834, this shrub was in flower on the 26th of January! [See Note *b*, p. 7.]

APRIL.

PHENOMENA.	Date of occurrence: Mean		Earliest	Latest
Rhyphus fenestralis, ap.	Apr. 4	10	Mar. 3	June 2
Black Poplar (*Populus nigra*), fl.	Apr. 4	2	Mar. 28	Apr. 11
Blackthorn (*Prunus spinosa*), fl.	Apr. 4	14	Mar. 14	Apr. 20
Bladder-nut (*Staphylea pinnata*), l.	Apr. 4	9	Mar. 16	Apr. 24
Crab (*Pyrus malus*), l.	Apr. 4	13	Mar. 15	Apr. 26
Lombardy Poplar (*P. pyramidalis*), fl.	Apr. 4	2	Apr. 2	Apr. 6
Six-cleft Plume-Moth (*Alucita hexadactyla*), ap. ...	Apr. 5	4	Feb. 3	May 5
Blackthorn (*Prunus spinosa*), l.	Apr. 5	10	Mar. 12	Apr. 14
Cherry (*Prunus cerasus*) l.	Apr. 5	8	Mar. 20	Apr. 20
Plum (*Prunus domestica*), l.	Apr. 5	10	Mar. 16	Apr. 22
Moorhen (*Gallinula chloropus*), lays	Apr. 6	3	Apr. 2	Apr. 8
Peacock Butterfly (*Vanessa io*), ap.	Apr. 6	12	Feb. 28	May 6
Turnip-fly (*Haltica nemorum*), ap.	Apr. 6	3	Mar. 25	Apr. 20
Rue-leaved Saxifrage (*Saxifraga tridactylites*), fl. ...	Apr. 6			
Chiff-chaff (*Sylvia hippolais*), note first heard ...	Apr. 7	12	Mar. 15	Apr. 24
Plum (*Prunus domestica*), fl.	Apr. 7	13	Mar. 7	Apr. 23
Blackbird (*Turdus merula*), lays	Apr. 8	7	Mar. 16	May 1
Fieldfares (*Turdus pilaris*), last seen	Apr. 8	9	Mar. 3	Apr. 30
Pheasant (*Phasianus colchicus*), utters its spring crow (a)	Apr. 8	8	Mar. 25	Apr. 19

Ringed Snake (*Natrix torquata*), comes abroad ...	Apr. 8	6	Mar. 25	Apr. 28
Rook (*Corvus frugilegus*), hatches	Apr. 8	17	Mar. 26	Apr. 23
Scarlet satin Mite (*Trombidium holosericeum*), ap. (b)	Apr. 8	3	Feb. 27	May 23
Laburnum (*Cytisus laburnum*), l.	Apr. 8	8	Mar. 25	Apr. 26
Larch (*Larix europæa*), fl.	Apr. 8	5	Apr. 1	Apr. 19
Nuthatch (*Sitta europæa*), whistling note heard ...	Apr. 9	3	Mar. 3	Apr. 29
Missel Thrush (*Turdus viscivorus*), lays	Apr. 9	3	Apr. 2	Apr. 15
Steropus madidus, ap.	Apr. 9	2	Mar. 29	Apr. 20
Common Elm (*Ulmus campestris*), l.	Apr. 9	17	Mar. 1	Apr. 26
Crown-Imperial (*Fritillaria imperialis*), fl.	Apr. 9	13	Mar. 13	Apr. 25
Pear (*Pyrus communis*), l.	Apr. 9	11	Mar. 20	Apr. 27
Red Currant (*Ribes rubrum*), fl.	Apr. 9	13	Mar. 28	Apr. 23
Turnip (*Brassica rapa*), fl.	Apr. 9	6	Mar. 12	May 5
Virginian Lungwort (*Mertensia pulmonarioides*), fl. ...	Apr. 10	6	Mar. 18	Apr. 27
Goldfinch (*Carduelis elegans*), sg. com.	Apr. 10	8	Mar. 1	May 1
Humble-bee Fly (*Bombylius medius*), ap.	Apr. 11	9	Apr. 1	May 1
Redbreast (*Erithaca rubecula*), lays	Apr. 11	3	Apr. 8	Apr. 14
Titlark (*Anthus pratensis*), sg. com.	Apr. 11	5	Mar. 29	Apr. 28
Alder (*Alnus glutinosa*), l.	Apr. 11	10	Mar. 19	Apr. 23

(a) This is in reference to the peculiar crow of the cock bird, when under the influence of sexual desire, and is only heard at the approach of, and during the breeding season: it is a good prognostic of warm spring weather.

(b) This mite, one of the largest and most conspicuous of its tribe, may be frequently observed in gardens, crawling upon the flower-beds, in the early spring.

APRIL.

Phenomena.	Date of occurrence: Mean		Earliest	Latest
Ash (*Fraxinus excelsior*), fl.	Apr. 11	16	Mar. 25	Apr. 27
Heart's-ease (*Viola tricolor*), fl.	Apr. 11	6	Mar. 26	Apr. 19
Common Lizard (*Zootoca vivipara*), ap.	Apr. 12	6	Mar. 12	Apr. 30
Song Thrush (*Turdus musicus*), lays	Apr. 12	6	Mar. 21	May 12
Birch (*Betula alba*), l.	Apr. 12	16	Mar. 23	May 4
Butter-bur (*Petasites vulgaris*), fl.	Apr. 12	6	Mar. 26	Apr. 29
Carrion-Beetle (*Necrophorus humator*), ap.	Apr. 13	4	Apr. 2	Apr. 22
Great Plover (*Œdicnemus crepitans*), first heard or seen	Apr. 13	4	Apr. 2	May 2
Stock-dove (*Columba œnas*), lays	Apr. 13	3	Mar. 27	Apr. 26
Pear (*Pyrus communis*), fl.	Apr. 13	12	Mar. 19	Apr. 24
Wild Guelder-rose (*Viburnum opulus*), l. (a)	Apr. 13	9	Apr. 4	Apr. 22
Chaffinch (*Fringilla cœlebs*), builds	Apr. 14	3	Apr. 10	Apr. 16
Redbreast (*Erithaca rubecula*), hatches	Apr. 14	3	Mar. 31	Apr. 22
Cherry (*Prunus cerasus*), fl.	Apr. 14	10	Mar. 12	Apr. 29
Evergreen Alkanet (*Anchusa sempervirens*), fl. ...	Apr. 14	6	Mar. 5	May 14
Hornbeam (*Carpinus betulus*), l.	Apr. 14	14	Mar. 28	Apr. 28
Lime (*Tilia europœa*), l.	Apr. 14	17	Mar. 27	Apr. 29
Pale Narcissus (*Narcissus biflorus*), fl.	Apr. 14	4	Apr. 1	May 2
Sycamore (*Acer pseudoplatanus*), l.	Apr. 14	17	Mar. 28	Apr. 26

Wild Tulip (*Tulipa sylvestris*), fl.	Apr. 14	3	Apr. 10	Apr. 23
Wren (*Troglodytes europæus*), builds	Apr. 14	3	Apr. 4	Apr. 21
Blackcap (*Curruca atricapilla*), first heard	Apr. 15	17	Mar. 28	May 1
Field Wood-rush (*Luzula campestris*), fl.	Apr. 15	12	Mar. 26	May 5
Strawberry-leaved Cinquefoil (*Potentilla fragariastrum*), fl.	Apr. 15	3	Mar. 21	Apr. 29
Long-eared Owl (*Otus vulgaris*), lays	Apr. 16			
Oiceoptoma thoracica, ap.	Apr. 16	3	Apr. 14	Apr. 18
Silpha obscura, ap.	Apr. 16			
Willow Warbler (*Sylvia trochilus*), first heard ([b]) ...	Apr. 16	17	Apr. 5	Apr. 25
Dog Violet (*Viola canina*), fl.	Apr. 16	7	Apr. 2	May 3
Maple (*Acer campestre*), l.	Apr. 16	17	Mar. 25	May 1
Frog Tadpoles, hatched	Apr. 17	3	Apr. 12	Apr. 22
Redstart (*Phœnicura ruticilla*), first heard	Apr. 17	17	Apr. 6	May 4
Small-white Butterfly (*Pontia rapæ*), ap.	Apr. 17	11	Mar. 14	May 12
Trailing Daphne (*Daphne cneorum*), fl.	Apr. 17	5	Apr. 14	May 3
Hedge Sparrow (*Accentor modularis*), hatches ...	Apr. 18			
Long-tailed Titmouse (*P. caudatus*), lays	Apr. 18			

([a]) In noting the periods of leafing and flowering in this shrub, it will be well in all cases to distinguish the wild sort from the large snowball variety cultivated in gardens.

([b]) In Mr Blomefield's original Calendar Apr. 15 was given from 12 years' observation as the mean date of the willow-wren's being first heard: he says that the same date was given by 21 years' observation; and adds that his date for the redstart (Apr. 15) and Apr. 20 for the tree pipit, may be considered as the means of 16 years: his date for the blackcap (Apr. 16) is the mean of 19 years, and Apr. 21 for the nightingale is the mean of 20 years. The dates for these birds in the present Calendar are but little changed.

APRIL.

Phenomena.	Date of occurrence: Mean		Earliest	Latest
Marsh Titmouse (*Parus palustris*), note ceases ...	Apr. 18			
Song Thrush (*Turdus musicus*), hatches	Apr. 18	3	Apr. 6	May 6
Fritillary (*Fritillaria meleagris*), fl.	Apr. 18	4	Apr. 9	Apr. 28
Hornbeam (*Carpinus betulus*), fl.	Apr. 18	8	Apr. 1	Apr. 30
White Poplar (*Populus alba*), l.	Apr. 18	11	Apr. 6	May 2
Wild Chervil (*Chærophyllum sylvestre*), fl.	Apr. 18	13	Mar. 27	May 4
Blackbird (*Turdus merula*), hatches	Apr. 19	4	Apr. 9	May 13
Black Slug (*Arion ater*), ap.	Apr. 19	3	Apr. 16	Apr. 26
Large-white Butterfly (*Pontia brassicæ*), ap.	Apr. 19	12	Mar. 16	May 11
Pœcilus cupreus, ap.	Apr. 19	5	Mar. 10	June 3
Queen Wasp (*Vespa vulgaris*), ap. ([a])	Apr. 19	11	Mar. 3	May 14
Sparkler-Beetle (*Cicindela campestris*), ap.	Apr. 19	5	Mar. 15	May 6
Corn Horsetail (*Equisetum arvense*), fl.	Apr. 19	3	Apr. 12	May 3
Meadow Lady's-smock (*Cardamine pratensis*), fl. ...	Apr. 19	15	Mar. 21	May 7
Plume-footed Bee (*Anthophora retusa*), ap. ([b]) ...	Apr. 19	3	Apr. 15	Apr. 23
Common Snail (*Helix aspersa*), comes abroad ...	Apr. 20	10	Mar. 27	May 25
Missel Thrush (*T. viscivorus*), hatches	Apr. 20	2	Apr. 10	Apr. 30
Black Currant (*Ribes nigrum*), fl.	Apr. 20	7	Apr. 6	Apr. 30
Nightingale (*Philomela luscinia*), first heard ...	Apr. 21	17	Apr. 8	Apr. 28
Swallow (*Hirundo rustica*), first seen ([c])	Apr. 21	17	Apr. 9	Apr. 28

Tree Pipit (*Anthus arboreus*), first heard	Apr. 21	17	Apr. 7	May 3
Birch (*Betula alba*), fl.	Apr. 21	9	Apr. 10	May 1
Narrow-leaved Mouse-ear-chickweed (*Cerast. triviale*), fl.	Apr. 21	5	Feb. 28	May 5
Wood Anemony (*Anemone nemorosa*), fl.	Apr. 21	6	Apr. 5	May 3
Wood Crowfoot (*Ranunculus auricomus*), fl.	Apr. 21	8	Mar. 31	May 5
Kestrel (*Falco tinnunculus*), lays	Apr. 22			
Ringed Snake (*Natrix torquata*), couples	Apr. 22			
Jelly Nostoc (*Nostoc commune*), ap. on lawns ...	Apr. 22	2	Apr. 19	Apr. 25
Jack-by-the-hedge (*Alliaria officinalis*), fl.	Apr. 22	16	Apr. 10	May 7
Peewit (*Vanellus cristatus*), lays (d)	Apr. 23	6	Mar. 29	May 20
Squirrel (*Sciurus vulgaris*), builds	Apr. 23			
Young Hedge Sparrows fledged	Apr. 23			
Young Moorhens hatched	Apr. 23			
Morell (*Morchella esculenta*), ap.	Apr. 23	4	Apr. 3	May 8
Black Poplar (*Populus nigra*), l.	Apr. 23			
Dogwood (*Cornus sanguinea*), l.	Apr. 23	6	Apr. 9	May 8
Lombardy Poplar (*Populus pyramidalis*), l.	Apr. 23	7	Apr. 13	May 2
Pasque-flower (*Anemone pulsatilla*), fl.	Apr. 23	10	Apr. 1	May 6

(a) This entry is in reference to the large queen wasps which appear in spring, and which are the founders of new colonies. The workers which abound so in the summer and autumn are their offspring, and do not show themselves till much later.

(b) This species of bee is not uncommon in gardens at Swaffham Bulbeck at this season, and is a good prognostic. It may often be observed hovering over the blossoms of the stinking hellebore (*Helleborus fœtidus*), to which plant it seems much attached.

(c) The date for the first arrival of the swallow, taking the mean of *twenty* years, is Apr. 19.

(d) It is this species which produces the "Plovers' eggs" eaten at table.

APRIL

Phenomena.	Date of occurrence Mean		Earliest	Latest
Jackdaw (*Corvus monedula*), lays	Apr. 23	3	Apr. 20	Apr. 27
Swallow (*Hirundo rustica*), sg. com.	Apr. 24	9	Apr. 16	May 5
Linnet (*Linota cannabina*), lays	Apr. 24	5	Apr. 2	May 10
Buttercup (*Ranunculus bulbosus*), fl.	Apr. 24	17	Apr. 1	May 5
Water Crowfoot (*Ranunculus aquatilis*), fl.	Apr. 24	7	Apr. 12	May 5
Ring-dove (*Columba palumbus*), lays	Apr. 25	4	Apr. 18	Apr. 30
Wryneck (*Yunx torquilla*), first heard	Apr. 25	12	Apr. 8	May 13
Beech (*Fagus sylvatica*), l.	Apr. 25	17	Apr. 10	May 6
Blue-bell (*Scilla nutans*), fl.	Apr. 25	2	Apr. 11	May 10
Green-veined-white Butterfly (*Pontia napi*), ap. ...	Apr. 25	7	Apr. 15	May 9
Common Snail (*Helix aspersa*), engenders	Apr. 26			
Lesser Whitethroat (*Curruca sylviella*), first heard ...	Apr. 26	17	Apr. 17	May 2
American Cowslip (*Dodecatheon meadia*), fl.	Apr. 26			
Oiceoptoma rugosa, ap.	Apr. 26	2	Apr. 13	May 10
Henbit Dead-nettle (*Lamium amplexicaule*), fl. ...	Apr. 26	6	Apr. 6	May 22
Strawberry (*Fragaria vesca*), fl.	Apr. 26	16	Apr. 10	May 13
Sedge Warbler (*Salicaria phragmitis*), first heard ...	Apr. 27	17	Apr. 15	May 9
Whitethroat (*Curruca cinerea*), first heard	Apr. 27	16	Apr. 14	May 10
Poplar Hawkmoth (*Smerinthus populi*), ap.	Apr. 27	2	Apr. 13	May 12
Ribwort Plantain (*Plantago lanceolata*), fl.	Apr. 27	12	Apr. 15	May 13

Chaffinch (*Fringilla cœlebs*), lays	Apr. 28	6	Apr. 17	May 4
Reed Bunting (*Emberiza schœniclus*), sg. com. ...	Apr. 28			
Germander Speedwell (*Veronica chamædrys*), fl. ...	Apr. 28	17	Apr. 9	May 13
Quince (*Cydonia vulgaris*), fl.	Apr. 28	5	Apr. 15	May 18
Banded Snail (*Helix nemoralis*), engenders	Apr. 29	2	Apr. 17	May 12
Cuckoo (*Cuculus canorus*), first heard	Apr. 29	17	Apr. 21	May 8
Large-tortoiseshell Butterfly (*Vanessa polychloros*), ap.	Apr. 29			
Crab (*Pyrus malus*), fl.	Apr. 29	13	Apr. 19	May 13
Maple (*Acer campestre*), fl.	Apr. 29	14	Apr. 17	May 13
Sycamore (*A. pseudoplatanus*), fl.	Apr. 29	13	Apr. 18	May 6
Walnut (*Juglans regia*), l.	Apr. 29	17	Apr. 14	May 14
Whinchat (*Saxicola rubetra*), first heard	Apr. 29	10	Apr. 18	May 31
Young Redbreasts fledged	Apr. 29	3	Apr. 22	May 4
Dung of cattle swarms with coleopterous insects ([a])	Apr. 30	3	Apr. 28	May 2
Lime Hawkmoth (*Smerinthus tiliæ*), ap.	Apr. 30	3	Apr. 23	May 10
Ring-dove (*Columba palumbus*), hatches	Apr. 30			
Speckled-wood Butterfly (*Hipparchia ægeria*), ap. ...	Apr. 30	6	Apr. 20	May 10
Young Song Thrushes fledged	Apr. 30	2	Apr. 17	May 13
Barberry (*Berberis vulgaris*), fl.	Apr. 30	13	Apr. 17	May 15
Fig (*Ficus carica*), l.	Apr. 30	9	Apr. 10	May 18
Wych Elm (*Ulmus montana*), l.	Apr. 30	5	Apr. 23	May 5

([a]) **These consist principally of *Aphodii* (*A. fossor, luridus, fimetarius,* and others) and the smaller *Staphylinidæ*.**

MAY.

Phenomena.	Date of occurrence			
	Mean		Earliest	Latest
Hister unicolor, ap.	May 1	2	Apr. 29	May 4
Woodruff (*Asperula odorata*), fl.	May 1	8	Apr. 15	May 12
Cuckow-pint (*Arum maculatum*), fl.	May 1	14	Apr. 10	May 12
Rhingia rostrata, ap.	May 2			
Apple tree, fl.	May 2	3	Apr. 27	May 9
Beech (*Fagus sylvatica*), fl.	May 2	6	Apr. 26	May 11
Black Medick (*Medicago lupulina*), fl.	May 2	12	Apr. 15	June 1
Caper Spurge (*Euphorbia lathyris*), fl.	May 2	5	Apr. 8	May 25
Dove's-foot Cranesbill (*Geranium molle*), fl.	May 2	10	Apr. 12	May 16
German Iris (*Iris germanica*), fl.	May 2	5	Apr. 19	May 24
Mealy-tree (*Viburnum lantana*), fl.	May 2	12	Apr. 24	May 13
Ranunculus acris, fl. pl. fl.	May 2	4	Apr. 21	May 12
Yellow Fumitory (*Corydalis lutea*), fl.	May 2	5	Apr. 14	May 17
Chaffinch (*Fringilla cœlebs*), hatches	May 3	2	May 3	May 4
Martin (*Hirundo urbica*), first seen	May 3	17	Apr. 15	May 24
Young Rooks fledged	May 3	15	Apr. 26	May 15
Silpha lævigata, ap.	May 3	3	Apr. 16	June 3
Wood Warbler (*Sylvia sibilatrix*), first heard	May 3	5	Apr. 27	May 7
Yellow Wagtail (*Motacilla raii*), first seen	May 3	6	Apr. 28	May 7
Ash (*Fraxinus excelsior*), l.	May 3	17	Apr. 22	May 12

Blue Sherardia (*Sherardia arvensis*), fl.	May 3	6	Mar. 11	May 28
Bugle (*Ajuga reptans*), fl.	May 3	14	Apr. 2	May 15
Lilac (*Syringa vulgaris*), fl.	May 3	15	Apr. 21	May 15
Medlar (*Mespilus germanica*), fl.	May 3			
Mullein Moth (*Cucullia verbasci*), ap.	May 4	2	Apr. 27	May 11
Bladder-nut (*Staphylea pinnata*), fl.	May 4	9	Apr. 20	May 14
Common Elm sheds its seed	May 4	4	Apr. 25	May 11
Field Chickweed (*Cerastium arvense*), fl.	May 4	9	Apr. 15	May 17
Young Blackbirds fledged	May 5	4	Apr. 26	May 18
Vine (*Vitis vinifera*), l.	May 4	8	Apr. 24	May 11
Dotterel makes its spring passage	May 5	4	Apr. 19	May 28
Empis pennipes ap.	May 5			
Latticed-heath Moth (*Hercyna clathrata*), ap. ...	May 5			
Sialis lutarius, ap.	May 5	8	Apr. 17	May 25
Horse-chestnut (*Æsculus hippocastanum*), fl. ...	May 5	16	Apr. 19	May 17
Oak (*Quercus robur*), l.	May 5	14	Apr. 25	May 17
Creeper (*Certhia familiaris*), builds	May 6	3	Apr. 13	May 25
Greenfinch (*Coccothraustes chloris*), builds	May 6			
Pheasant (*Phasianus colchicus*), lays	May 6	4	Apr. 23	May 20
Pale perfoliate Honeysuckle (*Lonicera caprifolium*), fl.	May 6	15	Mar. 14	May 19
Creeper (*Certhia familiaris*), lays	May 7	2	Apr. 26	May 18
Garden Warbler (*Curruca hortensis*), first heard ...	May 7	13	May 1	May 29
Long-eared Bat (*Plecotus auritus*), comes abroad ...	May 7			

MAY.

PHENOMENA.	Date of occurrence: Mean		Earliest	Latest
Sand-Martin (*H. riparia*), first seen	May 7	6	Apr. 20	May 23
Swallow (*Hirundo rustica*), builds	May 7			
Bitter Winter-cress (*Barbarea vulgaris*), fl.	May 7	9	Apr. 27	May 19
Herb-Robert (*Geranium robertianum*), fl.	May 7	12	Mar. 23	May 24
Mountain-ash (*Pyrus aucuparia*), fl.	May 7			
Snowdrop Anemony (*Anemone sylvestris*), fl.	May 7	5	Apr. 25	May 20
Whitethorn (*Cratægus oxyacantha*), fl.	May 7	17	Apr. 19	May 20
Emperor Moth (i.e. *Saturnia pavonia-minor*), ap.	May 8	3	Apr. 17	June 4
Greenfinch (*Coccothraustes chloris*), lays	May 8, 9	2		
Turtle-dove (*Columba turtur*), first heard	May 8	14	Apr. 27	May 20
Great Carex (*Carex riparia*), fl.	May 8	3	May 1	May 12
Plane (*Platanus orientalis*), l.	May 8	10	Apr. 29	May 22
Red Clover (*Trifolium pratense*), fl.	May 8	15	Apr. 25	May 19
Small marsh Valerian (*Valeriana dioica*), fl.	May 8	7	Apr. 20	May 28
Burying-Beetle (*Necrophorus vespillo*), ap.	May 9	3	Apr. 24	June 2
Flesh-fly (*Sarcophaga carnaria*), ap.	May 9	3	May 7	May 12
Persian Lilac (*Syringa persica*), fl.	May 9	7	Apr. 28	May 19
Thyme-leaved Speedwell (*Veronica serpyllifolia*), fl. ...	May 9	7	May 3	May 14
White Jasmine (*Jasminum officinale*), l.	May 9	4	May 3	May 15

Buff-tip Moth (*Pygæra bucephala*), ap.	May 10	2	May 7	May 13
Noctule Bat (*Vespertilio noctula*), comes abroad ...	May 10			
Pale tussock Moth (*Dasychira pudibunda*), ap. ...	May 10			
Partridge (*Perdix cinerea*), lays	May 10	2	May 6	May 14
London-pride (*Saxifraga umbrosa*), fl.	May 10	5	Apr. 29	May 29
Veronica gentianoides, fl.	May 10	3	May 3	May 17
Orange-tip Butterfly (*Pontia cardamines*), ap. ...	May 11	16	May 2	June 2
Reed Bunting (*Emberiza schæniclus*), lays	May 11			
Comfrey (*Symphytum officinale*), fl.	May 11	8	May 3	May 22
Common Fumitory (*Fumaria officinalis*), fl.	May 11	11	Apr. 25	June 12
Hemlock Stork's-bill (*Erodium cicutarium*), fl. ...	May 11	3	Apr. 16	June 20
Water-violet (*Hottonia palustris*), fl.	May 11	8	May 1	May 30
Laburnum (*Cytisus laburnum*), fl.	May 11	11	May 3	May 24
Harry-long-legs (*Tipula oleracea*), ap.	May 12	4	May 6	May 22
Long-tailed Titmouse (*Parus caudatus*), hatches ...	May 12			
Black Bog-rush (*Schænus nigricans*), fl.	May 12			
Corn Gromwell (*Lithospermum arvense*), fl.	May 12	4	May 5	May 17
Green-winged meadow Orchis (*Orchis morio*), fl. ...	May 12	3	May 9	May 19
Lesser Burnet (*Poterium sanguisorba*), fl.	May 12	8	Apr. 30	June 3
Oak (*Quercus robur*), fl.	May 12	10	Apr. 30	June 2
Wild Sage (*Salvia verbenacea*), fl.	May 12	16	May 1	June 1
Nightingale (*Philomela luscinia*), lays	May 13	2	May 8	May 18
Scorpion-fly (*Panorpa communis*), ap.	May 13	4	May 11	May 25

MAY.

PHENOMENA.	Date of occurrence			
	Mean		Earliest	Latest
Willow Warbler (*Sylvia trochilus*), lays	May 13			
Aspen (*Populus tremula*), l.	May 13	5	May 3	May 20
Celandine (*Chelidonium majus*), fl.	May 13	11	May 1	May 25
Celery-leaved Crowfoot (*Ranunculus sceleratus*), fl. ...	May 13	5	Apr. 29	June 4
Charlock (*Brassica sinapistrum*), fl.	May 13	7	May 2	May 23
Corn-salad (*Valerianella olitoria*), fl.	May 13	2	May 5	May 21
Lily-of-the-valley (*Convallaria majalis*), fl.	May 13	9	Apr. 29	May 29
Walnut (*Juglans regia*), fl.	May 13	9	May 2	May 27
Yellow Archangel (*Lamium galeobdolon*), fl.	May 13	2	May 8	May 18
Dingy-skipper Butterfly (*Thymele tages*), ap.	May 14			
Dot Moth (*Mamestra persicariæ*), ap.	May 14	2	May 10	May 18
Swift (*Cypselus apus*), first seen	May 14	13	May 6	May 30
Tit Lark (*Anthus pratensis*), lays	May 14			
Fly Orchis (*Ophrys muscifera*), fl.	May 14	4	May 11	May 21
Meadow Fox-tail-grass (*Alopecurus pratensis*), fl. ...	May 14	8	May 5	June 2
Shepherd's-needle (*Scandix pecten-veneris*), fl. ...	May 14	5	May 8	May 31
Sweet-scented Vernal-grass (*Anthoxanthum odoratum*), fl.	May 14	2	May 12	May 17
Water Avens (*Geum rivale*), fl.	May 14	2	May 9	May 19
Yellow-hammer (*Emberiza citrinella*), lays	May 15	3	May 4	May 22

Columbine (*Aquilegia vulgaris*), fl.	May 15	8	May 8	May 28
Milk-wort (*Polygala vulgaris*), fl.	May 15	8	Apr. 20	May 31
Solomon's-seal (*Polygonatum multiflorum*), fl. ...	May 15	3	May 7	May 21
Lesser Whitethroat (*Curruca sylviella*), lays	May 16			
Harpalus ruficornis, ap.	May 16	2	May 4	May 29
Common Tormentil (*Potentilla tormentilla*), fl. ...	May 16	5	May 1	May 28
Wall Butterfly (*Hipparchia megæra*), ap.	May 17	9	May 4	June 20
Young broods of Chaffinches fledged	May 17			
Alder Buckthorn (*Rhamnus frangula*), fl.	May 17			
Common Hedge-mustard (*Sisymbrium officinale*), fl. ...	May 17	7	Apr. 30	June 3
Holly (*Ilex aquifolium*), fl.	May 17	7	May 5	June 10
Tway-blade (*Listera ovata*), fl.	May 17	5	May 14	May 21
Goldfinch (*Carduelis elegans*), lays	May 18			
Spotted Flycatcher (*Muscicapa grisola*), first seen ...	May 18	12	May 12	May 31
Tree Pipit (*Anthus arboreus*), lays	May 18			
Buck-bean (*Menyanthes trifoliata*), fl.	May 18			
Field Flea-wort (*Senecio campestris*), fl.	May 18	3	May 3	June 7
Field Scorpion-grass (*Myosotis arvensis*), fl.	May 18	6	May 7	June 2
Silver-weed (*Potentilla anserina*), fl.	May 18	10	Apr. 30	June 4
Star-of-Bethlehem (*Ornithogalum umbellatum*), fl. ...	May 18	3	May 14	May 26
Upright Crowfoot (*R. acris*), fl.	May 18	8	May 4	May 30
Blackcap (*Curruca atricapilla*), lays	May 19			
Young broods of Greenfinches hatched	May 19	3	May 13	May 28

MAY.

PHENOMENA.	Date of occurrence Mean		Earliest	Latest
Young broods of Starlings fledged	May 19			
May-fly (*Ephemera vulgata*), ap.	May 19	11	Apr. 29	June 9
Mulberry (*Morus nigra*), l.	May 19	4	May 12	May 28
Pæony (*Pæonia officinalis*), fl.	May 19	9	May 6	June 2
Ragged-robin (*Lychnis flos-cuculi*), fl.	May 19	6	May 10	June 9
Sedge Warbler (*Salicaria phragmitis*), lays	May 20			
Whinchat (*Saxicola rubetra*), lays	May 20			
Cross-wort (*Galium cruciatum*), fl.	May 20			
Cockchaffer (*Melolontha vulgaris*), ap.	May 21	11	Apr. 14	June 13
Golden-green Dragon-fly (*Calepteryx virgo*), ap. ...	May 21			
Malachius æneus, ap.	May 21			
White Campion (*Lychnis vespertina*), fl.	May 21	9	May 6	June 4
Great Titmouse (*Parus major*), sg. ceas.	May 22	9	May 5	June 14
House-martin (*Hirundo urbica*), builds	May 22	7	May 11	May 31
Greasy-Fritillary Butterfly (*Melitæa artemis*), ap. ...	May 23			
Midge (*Thrips physaphus*), ap.	May 23	9	Mar. 27	July 12
Soldier-Beetle (*Telephorus lividus*), ap.	May 23	6	May 9	June 3
Common Gromwell (*Lithospermum officinale*), fl. ...	May 23	5	May 16	May 29
Raspberry (*Rubus idæus*), fl.	May 23	11	May 3	June 8
White Clover (*Trifolium repens*), fl.	May 23	8	May 3	June 9

Small-heath Butterfly (*Hipparchia pamphilus*), ap. ...	May 24	12	May 14	June 11
Heath Moth (*Fidonia atomaria*), ap.	May 24	2	May 19	May 29
Sea Pancratium (*Pancratium maritimum*), fl. ...	May 24	6	May 16	June 2
Slender Fox-tail-grass (*Alopecurus agrestis*), fl. ...	May 24	6	Apr. 30	June 22
Symphytum asperrimum, fl.	May 24	3	May 19	May 30
Wild Mignonette (*Reseda lutea*), fl.	May 24	6	May 4	June 8
Wild Radish (*Rhaphanus raphanistrum*), fl.	May 24	5	May 2	June 6
Bistort (*Polygonum bistorta*), fl.	May 25	4	May 8	June 22
Brooklime (*Veronica beccabunga*), fl.	May 25	8	May 13	June 11
Butterwort (*Pinguicula vulgaris*), fl.	May 25			
Dwarf dark-winged Orchis (*Orchis ustulata*), fl. ...	May 25	7	May 3	June 3
Guelder-rose (*Viburnum opulus*), fl.	May 25	14	May 12	June 9
Herb-Bennet (*Geum urbanum*), fl.	May 25	12	May 14	June 6
Mouse-ear Hawkweed (*Hieracium pilosella*), fl. ...	May 25	7	May 16	June 7
Ox-eye-Daisy (*Chrysanthemum leucanthemum*), fl. ...	May 25	14	May 12	June 11
Purple mountain Milk-vetch (*Astragalus hypoglottis*), fl.	May 25	6	May 3	June 7
Soft Brome-grass (*Bromus mollis*), fl.	May 25	3	May 21	June 2
Brown-argus Butterfly (*Polyommatus agrestis*), ap. ...	May 26	2	May 19	June 3
Grizzled-Skipper Butterfly (*Thymele alveolus*), ap. ...	May 26	4	May 18	June 13
Stinging-fly (*Stomoxys calcitrans*), ap.	May 26	8	May 8	June 19
Bird's-foot Trefoil (*Lotus corniculatus*), fl.	May 26	6	May 1	June 3
Early purple Orchis (*Orchis mascula*), fl.	May 26	3	May 9	June 20
Monk's-hood (*Aconitum napellus*), fl.	May 26	5	May 8	June 27

MAY

PHENOMENA.	Date of occurrence			
	Mean		Earliest	Latest
Garden Carpet-Moth (*Cidaria fluctuata*), ap.	May 27			
Missel Thrush (*Turdus viscivorus*), sg. ceas.	May 27	7	May 3	June 24
Common Rock-rose (*Helianthemum vulgare*), fl. ...	May 27	7	May 16	June 7
Common Sorrel (*Rumex acetosa*), fl.	May 27	8	May 18	June 14
Evergreen Oak (*Quercus ilex*), fl.	May 27	2	May 7	June 16
Hoary Plantain (*Plantago media*), fl.	May 27	6	May 18	June 9
Tufted Horse-shoe-vetch (*Hippocrepis comosa*), fl. ...	May 27	6	May 9	June 3
Wood Scorpion-grass (*Myosotis sylvatica*), fl.	May 27			
Anthomyia canicularis, ap.	May 28			
Common Sandpiper (*Totanus hypoleucos*), first seen ...	May 28			
Hive Bee swarms	May 28	7	May 14	June 13
Sailor-Beetle (*Telephorus rusticus*), ap.	May 28	5	May 20	June 2
Dewberry (*Rubus cæsius*), fl.	May 28	6	May 16	June 13
Hound's-tongue (*Cynoglossum officinale*), fl.	May 28	6	May 13	June 20
Red Bryony (*Bryonia dioica*), fl.	May 28	12	May 11	June 9
Smooth-stalked Meadow-grass (*Poa pratensis*), fl. ...	May 28			
Wood Sanicle (*Sanicula europæa*), fl.	May 28			
Landrail (*Crex pratensis*), note first heard	May 29	4	May 4	July 16
Goose-grass (*Galium aparine*), fl.	May 29	10	May 9	July 1
Libellula depressa, ap.	May 30	5	May 13	June 15

Quail (*Coturnix vulgaris*), note first heard	May 30	6	May 1	July 16
Common Garden Pink, fl.	May 30	3	May 28	June 3
Corn Crowfoot (*Ranunculus arvensis*), fl.	May 30	4	May 19	June 5
Hairy Thrincia (*Leontodon hirtus*), fl.	May 30	2	May 25	June 4
Marsh Lousewort (*Pedicularis palustris*), fl.	May 30	4	May 21	June 7
Syringa (*Philadelphus coronarius*), fl.	May 30	13	May 19	June 10
Yellow Rattle (*Rhinanthus crista-galli*), fl.	May 30	5	May 20	June 12
Puss Moth (*Cerura vinula*), ap.	May 31	2	May 21	June 10
Young broods of Whitethroats (*Curruca cinerea*), fledged	May 31	2	May 29	June 3
Buckthorn (*Rhamnus catharticus*), fl.	May 31	5	May 21	June 9
Common Elder (*Sambucus nigra*), fl.	May 31	17	May 17	June 13
Flat-stalked Meadow-grass (*Poa compressa*), fl. ...	May 31			
Marsh Orchis (*Orchis latifolia*), fl.	May 31	5	May 5	June 28
Small Nettle (*Urtica urens*), fl.	May 31	5	Apr. 9	July 19
Yellow Day-lily (*Hemerocallis flava*), fl.	May 31	8	May 25	June 9

JUNE.

Young broods of Linnets fledged	June 1	2	May 28	June 6
Swallow-tail Butterfly (*Papilio machaon*), ap. ...	June 1	10	May 11	June 28
Burnet Rose (*Rosa spinosissima*), fl.	June 1	6	May 20	June 12
Papaver bracteatum, fl.	June 1	2	May 28	June 6
Rough Chervil (*Chærophyllum temulum*), fl.	June 1	8	May 15	June 13

JUNE.

PHENOMENA.	Date of occurrence: Mean	Earliest	Latest
Rye (*Secale cereale*), fl.	June 1 4	May 26	June 7
Four-spotted Dragon-fly (*Libellula quadrimaculata*), ap.	June 2 7	May 17	July 4
Italian Bugloss (*Anchusa italica*), fl.	June 2 4	May 23	June 12
Snowberry (*Symphoricarp. racem.*), fl.	June 2 3	May 26	June 7
Virginian Spider-wort (*Tradescantia virginica*), fl. ...	June 2 6	May 24	June 13
Small garden Chaffer (*Anomala horticola*), ap. ...	June 3 3	May 30	June 11
Deadly Nightshade (*Atropa belladonna*), fl.	June 3 8	May 20	June 13
Long-prickly-headed Poppy (*Papaver argemone*), fl. ...	June 3 5	May 23	June 29
Thyme-leaved Sandwort (*Arenaria serpyllifolia*), fl. ...	June 3 6	May 13	July 10
Common-blue Butterfly (*Polyommatus alexis*), ap. ...	June 4 10	May 19	June 28
Spotted Flycatcher (*Muscicapa grisola*), lays ...	June 4 4	May 24	June 14
Young broods of Pheasants hatched	June 4 4	May 28	June 18
Pyrochroa rubens, ap.	June 4 7	May 17	July 14
Redbreast lays a second time	June 4 2	May 29	June 9
Clustered Bell-flower (*Campanula glomerata*), fl. ...	June 4 6	May 3	June 26
Common red Poppy (*Papaver rhœas*), fl.	June 4 8	May 22	June 15
Downy Oat-grass (*Avena pubescens*), fl.	June 4 2	May 28	June 12
Rye-grass (*Lolium perenne*), fl.	June 4 6	May 20	June 15
Creophilus maxillosus, ap.	June 5 4	Apr. 17	July 23

Fraxinella (*Dictamnus alba*), fl.	June 5	5	May 25	June 16
Red Valerian (*Centranthus ruber*), fl.	June 5	3	May 29	June 9
Bladder-campion (*Silene inflata*), fl.	June 6	10	May 18	June 25
Great Nettle (*Urtica dioica*), fl.	June 6	6	May 28	June 16
Large Oriental Poppy (*Papaver orientale*), fl. ...	June 6	6	May 30	June 12
Spotted palmate Orchis (*Orchis maculata*), fl. ...	June 6			
Black Bryony (*Tamus communis*), fl.	June 7	6	May 28	June 12
Cock's-foot-grass (*Dactylis glomerata*), fl.	June 7	5	May 21	June 22
Flixweed (*Sisymbrium sophia*), fl.	June 7	6	Apr. 25	July 18
Mountain Cudweed (*Antennaria dioica*), fl.	June 7	4	May 14	June 29
Purging Flax (*Linum catharticum*), fl.	June 7	6	May 19	June 26
Common Speedwell (*Veronica officinalis*), fl.	June 8	2	June 6	June 10
Common Vetch (*Vicia sativa*), fl.	June 8	4	May 23	June 24
Dog-rose (*Rosa canina*), fl.	June 8	12	May 26	June 22
Jagged-leaved Crane's-bill (*Geranium dissectum*), fl. ...	June 8	4	June 1	June 20
Saintfoin (*Onobrychis sativa*), fl.	June 8	7	May 23	June 26
Scarlet Pimpernel (*Anagallis arvensis*), fl.	June 8	6	May 7	June 30
Nightingale (*Philomela luscinia*), sg. ceas.	June 9	11	May 24	June 20
Barren Brome-grass (*Bromus sterilis*), fl.	June 9	2	May 26	June 24
Bloody Crane's-bill (*Geranium sanguineum*), fl. ...	June 9	7	May 19	June 27
Dogwood (*Cornus sanguinea*), fl.	June 9	10	May 22	June 30
Dwarf Mallow (*Malva rotundifolia*), fl.	June 9	7	May 16	June 30
Small Bindweed (*Convolvulus arvensis*), fl.	June 9	10	May 25	June 28

JUNE.

PHENOMENA.	Date of occurrence: Mean		Earliest	Latest
Water-cress (*Nasturtium officinale*), fl.	June 9	7	May 12	July 15
Wild Thyme (*Thymus serpyllum*), fl.	June 9	10	May 23	June 27
Bright-line brown-eye Moth (*Mamestra oleracea*), ap.	June 10			
Large brown Dragon-fly (*Æshna grandis*), ap. ...	June 10	3	May 17	July 10
Spotted Flycatcher (*Muscicapa grisola*), hatches ...	June 10			
Butterfly Orchis (*Habenaria bifolia*), fl.	June 10			
Common Honeysuckle (*Lonicera periclymenum*), fl. ...	June 10	10	May 12	June 26
Cow-parsnep (*Heracleum sphondylium*), fl.	June 10	6	May 7	July 20
Hard Rush (*Juncus glaucus*), fl.	June 10			
Mulberry (*Morus nigra*), fl.	June 10	4	May 30	June 22
Yellow Flag (*Iris pseudacorus*), fl.	June 10	6	May 30	July 3
Young Jackdaws fledged	June 11			
Common Mallow (*M. sylvestris*), fl.	June 11	10	June 2	June 25
Great Snapdragon (*Antirrhinum majus*), fl.	June 11	9	May 25	July 1
Hedge Woundwort (*Stachys sylvatica*), fl.	June 11	6	June 1	June 29
Lesser Spearwort (*Ranunculus flammula*), fl.	June 11	4	May 27	June 20
Meadow Hay cut	June 11	17	May 30	June 25
Small Bugloss (*Lycopsis arvensis*), fl.	June 11	3	June 8	July 5
White Mustard (*Brassica alba*), fl.	June 11	4	May 23	June 25
Yellow Vetchling (*Lathyrus aphaca*), fl.	June 11	2	June 1	June 21

Second broods of Redbreasts hatched	June 12			
Small-blue Butterfly (*Polyommatus alsus*), ap. ...	June 12	4	June 3	June 19
Small elephant Hawkmoth (*Deilephila porcellus*), ap.	June 12			
Curled Dock (*Rumex crispus*), fl.	June 12	5	May 20	July 10
Gout-weed (*Ægopodium podagraria*), fl.	June 12	2	June 9	June 15
Jacob's-ladder (*Polemonium cæruleum*), fl.	June 12	5	May 7	July 7
Rough-stalked Meadow-grass (*Poa trivialis*), fl. ...	June 12	4	June 2	June 23
Smooth Hawk's-beard (*Crepis virens*), fl.	June 12	4	May 16	June 22
Redstart (*Phœnicura ruticilla*), sg. ceas.	June 13	9	May 27	July 11
Tissue Moth (*Triphosa dubitata*), ap.	June 13	3	Apr. 16	Sept. 19
Dyer's Rocket (*Reseda luteola*), fl.	June 13	7	May 25	July 10
Hazel-leaved Bramble (*Rubus corylifolius*), fl. ...	June 13	5	June 3	June 15
Officinal Sage (*Salvia officinalis*), fl.	June 13	3	May 27	June 29
Upright Brome-grass (*Bromus erectus*), fl.	June 13	3	June 11	June 15
Woody Nightshade (*Solanum dulcamara*), fl. ...	June 13	10	June 3	June 27
Landrail (*Crex pratensis*), lays	June 14			
Large-skipper Butterfly (*Pamphila sylvanus*), ap. ...	June 14	5	June 6	June 29
Pink-underwing Moth (*Callimorpha jacobææ*), ap. ...	June 14	4	May 19	July 4
Young broods of Swallows fledged	June 14			
Floating Meadow-grass (*Glyceria fluitans*), fl. ...	June 14	3	June 2	July 5
Fox-glove (*Digitalis purpurea*), fl.	June 14	4	June 6	July 2
Frog-bit (*Hydrocharis morsus-ranæ*), fl.	June 14			
Lady's-fingers (*Anthyllis vulneraria*), fl.	June 14	6	May 16	July 30

JUNE.

PHENOMENA.	Date of occurrence: Mean		Earliest	Latest
Long-rooted Cat's-ear (*Hypochæris maculata*), fl. ...	June 14	2	June 12	June 16
Meadow-rue (*Thalictrum flavum*), fl.	June 14	6	May 22	July 14
Moss Rose (*Rosa muscosa*), fl.	June 14			
Rosa centifolia, fl.	June 14	3	June 13	June 14
Rough Hawk-bit (*Leont[illegible]don hispidus*), fl.	June 14	6	May 26	July 16
Young broods of Redsta[illegible] edged	June 15	2	June 6	June 25
Common Wart-cress (*Senebiera coronopus*), fl. ...	June 15	5	June 9	June 21
Corn Blue-bottle (*Centaurea cyanus*), fl.	June 15	7	May 23	July 9
Creeping Cinque-foil (*Potentilla reptans*), fl.	June 15	6	June 10	June 20
Great Hedge Bedstraw (*Galium mollugo*), fl. ...	June 15	7	May 23	June 30
Henbane (*Hyoscyamus niger*), fl.	June 15	5	May 22	June 30
Mare's-tail (*Hippuris vulgaris*), fl.	June 15			
Melilot (*Melilotus officinalis*), fl.	June 15	4	June 4	June 26
Musk Thistle (*Carduus nutans*), fl. ([a])	June 15	7	May 24	July 16
Oat-grass (*Arrhenatherum avenaceum*), fl.	June 15	2	June 13	June 17
Quaking-grass (*Briza media*), fl.	June 15	4	June 2	June 23
Worm-seed Treacle-mustard (*Erysimum cheiranthoides*), fl.	June 15	7	Apr. 14	July 27
Dagger Moth (*Acronycta psi*), ap.	June 16	6	May 15	July 20
Corn Bell-flower (*Specularia hybrida*), fl.	June 16	4	May 24	June 30

Ivy casts its leaves	June 16	3	June 9	June 29
Narrow-leaved Oat-grass (*Avena pratensis*), fl. ...	June 16	2	June 12	June 20
Sow-thistle (*Sonchus oleraceus*), fl.	June 16	6	May 24	July 2
Trailing Dog-rose (*Rosa arvensis*), fl.	June 16	5	June 8	June 22
Wild Chamomile (*Matricaria chamomilla*), fl. ...	June 16			
Wild Parsnep (*Pastinaca sativa*), fl.	June 16	9	May 7	July 12
Good-King-Henry (*Chenopodium bonus-henricus*), fl. ...	June 17	4	May 31	Aug. 5
Meadow Crane's-bill (*Geranium pratense*), fl.	June 17	5	June 1	July 4
Milk-thistle (*Silybum marianum*), fl.	June 17			
Sweet-William (*Dianthus barbatus*), fl.	June 17	5	June 14	June 20
Frog tadpoles nearly full grown, and acquiring fore feet	June 18	2	June 17	June 20
Turtle-dove (*Columba turtur*), lays	June 18			
Corn-cockle (*Lychnis githago*), fl.	June 18	5	May 23	July 7
Creeping Spike-rush (*Eleocharis palustris*), fl. ...	June 18	3	June 11	June 30
Forget-me-not (*Myosotis palustris*), fl.	June 18	5	June 7	July 4
Sweet-briar (*Rosa rubiginosa*), fl.	June 18	3	June 13	June 27
Atopa cervina, ap.	June 19	3	June 14	June 26
Silver Y Moth (*Plusia gamma*), ap.	June 19	6	May 5	Aug. 2
Bee Orchis (*Ophrys apifera*), fl.	June 19	7	June 9	June 28
Pellitory-of-the-wall (*Parietaria officinalis*), fl. ...	June 19	6	May 9	July 20

([a]) Linnæus states in respect of the Thistles in general, "*Cardui varii non florent antequam solstitium absolutum est.*" (Philos. Bot.) This species, however, is an exception, and appears to be very uncertain in its time of flowering.

JUNE.

Phenomena.	Date of occurrence: Mean		Earliest	Latest
Water Speedwell (*Veronica anagallis*), fl.	June 19	5	June 7	July 9
Eyed Hawk-moth (*Smerinthus ocellatus*), ap.	June 20			
Young broods of Greenfinches fledged	June 20			
Biting Stonecrop (*Sedum acre*), fl.	June 20	8	June 9	July 12
Black Knapweed (*Centaurea nigra*), fl.	June 20	9	May 22	July 2
Small Scabious (*Scabiosa columbaria*), fl.	June 20	3	June 7	July 5
Strawberries ripe	June 20	10	June 2	July 5
Sweet Milk-vetch (*Astragalus glycyphyllos*), fl.	June 20			
Viper's Bugloss (*Echium vulgare*), fl.	June 20	5	June 10	June 30
Water Figwort (*Scrophularia aquatica*), fl.	June 20	6	May 24	July 4
Wild Carrot (*Daucus carota*), fl.	June 20	6	June 14	July 2
Asparagus-Beetle (*Crioceris asparagi*), ap.	June 21	2	June 2	July 10
Young broods of Partridges hatched	June 21			
Rose Beetle (*Cetonia aurata*), ap.	June 21	2	June 17	June 26
Common Cudweed (*Filago germanica*), fl.	June 21	5	June 3	July 15
Goat's-beard (*Tragopogon pratensis*), fl.	June 21	4	June 7	July 10
Marsh Thistle (*Carduus palustris*), fl.	June 21	4	June 14	June 28
Sweet-smelling Orchis (*Gymnadenia conopsea*), fl.	June 21	3	June 18	June 24
Tawny Day-lily (*Hemerocallis fulva*), fl.	June 21	5	June 6	July 6
Six-spot-Burnet Moth (*Anthrocera filipendulæ*), ap.	June 22	5	Apr. 20	July 19

Self-heal (*Prunella vulgaris*), fl.	June 22	7	June 9	June 30
Wheat (*Triticum hybernum*), fl.	June 22	13	June 9	July 11
Gold-crested Wren (*Regulus cristatus*), sg. ceas. ...	June 23	5	May 27	July 16
Redbreast (*Erithaca rubecula*), sg. ceas.	June 23	4	June 9	July 11
Broad-leaved Dock (*Rumex obtusifolius*), fl.	June 23	5	June 5	July 4
Crested Dog's-tail-grass (*Cynosurus cristatus*), fl. ...	June 23	2	June 17	June 29
Meadow Vetchling (*Lathyrus pratensis*), fl.	June 23	5	June 12	July 8
Nipplewort (*Lapsana communis*), fl.	June 23	5	June 13	June 30
Stinking Horehound (*Ballota nigra*), fl.	June 23	7	June 12	July 11
Painted-lady Butterfly (*Vanessa cardui*), ap.	June 24	2	June 20	June 28
Privet Hawkmoth (*Sphinx ligustri*), ap.	June 24	6	June 12	July 4
Enchanter's-Nightshade (*Circæa lutetiana*), fl. ...	June 24	6	June 10	July 4
Hop-Trefoil (*Trifolium procumbens*), fl.	June 24	2	June 8	July 10
Portugal Laurel (*Prunus lusitanica*), fl.	June 24, 25			
Turk's-cap Lily (*Lilium martagon*), fl.	June 24	3	June 18	June 27
Wild Oat (*Avena fatua*), fl.	June 24			
Small Horse-fly (*Hæmatopota pluvialis*), ap.	June 25	3	June 18	June 29
Bee Larkspur (*Delphinium elatior*), fl.	June 25	3	June 22	June 30
Chamomile (*Anthemis nobilis*), fl.	June 25			
Crested Hair-grass (*Koeleria cristata*), fl.	June 25			
Dropwort (*Spiræa filipendula*), fl.	June 25	5	June 17	July 9
May-weed (*Anthemis cotula*), fl.	June 25	6	June 11	July 12
Meadow-brown Butterfly (*Hipparchia janira*), ap. ...	June 26	8	June 9	July 10

JUNE.

Phenomena.	Date of occurrence: Mean		Earliest	Latest
Bull-rush (*Scirpus lacustris*), fl. ([a])	June 26	2	June 13	July 9
French Willow-herb (*Epilobium angustifolium*), fl. ...	June 26	3	June 26	June 27
Great Spearwort (*Ranunculus lingua*), fl.	June 26	6	June 14	July 10
Privet (*Ligustrum vulgare*), fl.	June 26	10	June 9	July 20
Spotted Cat's-ear (*Achyrophorus maculatus*), fl. ...	June 26	2	June 14	July 8
Welted Thistle (*Carduus acanthoides*), fl.	June 26	8	May 26	July 13
Yellow Water-lily (*Nuphar lutea*), fl.	June 26	5	May 30	July 26
Cuckoo (*Cuculus canorus*), last heard	June 27	10	June 6	July 11
White-plume Moth (*Pterophorus pentadactylus*), ap. ...	June 27			
Amphibious Nasturtium (*Nasturtium amphibium*), fl.	June 27	3	June 7	July 15
Cherries ripe	June 27			
Couch-grass (*Triticum repens*), fl.	June 27			
Meadow Soft-grass (*Holcus lanatus*), fl.	June 27	2	June 25	June 29
Orange Lily (*Lilium bulbiferum*), fl.	June 27	4	June 16	July 7
Spurge-laurel (*Daphne laureola*), berries ripe ...	June 27	3	June 16	July 3
Squinancy-wort (*Asperula cynanchica*), fl.	June 27	4	June 20	July 7
Wasp-Beetle (*Clytus arietis*), ap.	June 28	2	June 14	July 12
Wood Warbler (*Sylvia sibilatrix*), sg. ceas.	June 28	2	June 15	July 11
Great Plantain (*Plantago major*), fl.	June 28	5	June 17	July 8
Hairy St John's Wort (*Hypericum hirsutum*), fl. ...	June 28	7	June 14	July 7

Strawberry Trefoil (*Trifolium fragiferum*), fl. ...	June 28	6	May 28	July 20
Common Wasp (*Vespa vulgaris*), begins to abound ...	June 29	3	June 25	July 4
Flax-leaved Bastard-Toad-flax (*Thesium linophyllum*), fl.	June 29	2	June 18	July 10
Millefoil (*Achillea millefolium*), fl.	June 29	9	June 1	July 18
Water Chickweed (*Stellaria aquatica*), fl.	June 29	4	June 8	July 16
Great Horse-fly (*Tabanus bovinus*), ap.	June 30	3	June 17	July 16
Whinchat (*Saxicola rubetra*), sg. ceas.	June 30	5	June 16	July 10
Basil-Thyme (*Calamintha acinos*), fl.	June 30	2	June 20	July 10
Blackberry (*Rubus fruticosus*), fl.	June 30	11	June 5	July 17
Brook-weed (*Samolus valerandi*), fl.	June 30	3	June 28	July 5
Field Scabious (*Knautia arvensis*), fl.	June 30	11	June 14	July 19
Meadow Pepper-saxifrage (*Silaus pratensis*), fl. ...	June 30			
Meadow-sweet (*Spiræa ulmaria*), fl.	June 30	8	June 14	July 15
White water Bedstraw (*Galium palustre*), fl.	June 30	4	June 8	July 14
White Water-lily (*Nymphæa alba*), fl.	June 30	7	June 2	Aug. 2

JULY.

Omaloplia ruricola, ap.	July 1	2	June 26	July 5
Barley (*Hordeum vulgare*), fl.	July 1	7	June 17	July 13
Borage (*Borago officinalis*), fl.	July 1	3	June 20	July 11
Common Agrimony (*Agrimonia eupatoria*), fl. ...	July 1	7	June 20	July 20

([a]) [*Typha* is more usually known as the Bull-rush.]

JULY.

Phenomena	Date of occurrence: Mean		Earliest	Latest
Pyramidal Orchis (*Orchis pyramidalis*), fl.	July 1	5	June 22	July 9
Round-leaved Bell-flower (*Campanula rotundifolia*), fl.	July 1	8	May 27	July 27
Water-dropwort (*Œnanthe fistulosa*), fl.	July 1	3	June 7	July 17
Second broods of House-sparrows hatched	July 2			
Rooks return to their nest trees to roost (a)	July 2	7	June 25	July 15
Common Skull-cap (*Scutellaria galericulata*), fl. ...	July 2	5	June 14	July 19
Great wild Valerian (*Valeriana officinalis*), fl. ...	July 2	3	June 14	July 24
Lime (*Tilia europæa*), fl.	July 2	9	June 21	July 15
Ragwort (*Senecio jacobæa*), fl.	July 2	9	June 24	July 9
Raspberries ripe	July 2	6	June 14	July 13
Rest-harrow (*Ononis arvensis*), fl.	July 2	4	June 15	July 10
Rose Campion (*Lychnis coronaria*), fl.	July 2			
Smooth-leaved Willow-herb (*Epilobium montanum*), fl.	July 2	4	June 16	July 25
Yellow Toadflax (*Linaria vulgaris*), fl.	July 2	8	June 16	July 23
Young Frogs come on land	July 3	3	June 28	July 8
Corn Sowthistle (*Sonchus arvensis*), fl.	July 3	5	June 21	July 18
Field Larkspur (*Delphinium consolida*), fl.	July 3	6	June 8	July 26
Great Knapweed (*Centaurea scabiosa*), fl.	July 3	8	June 20	July 19
Red currants ripe	July 3	6	June 19	July 9
Tufted Vetch (*Vicia cracca*), fl.	July 3	4	June 15	July 20

Elephant Hawkmoth (*Deilephila elpenor*), ap. ...	July 4	4	June 12	July 29
Ghost Moth (*Hepialus humuli*), ap.	July 4	3	June 18	July 20
Young Jays (*Garrulus glandarius*) fledged	July 4			
Midsummer-Dor (*Melolontha solstitialis*), ap.	July 4	11	June 15	July 17
Wood-leopard Moth (*Zeuzera æsculi*), ap.	July 4			
Common St John's-wort (*Hypericum perforatum*), fl. ...	July 4	8	June 27	July 6
Dark Mullein (*Verbascum nigrum*), fl.	July 4	4	June 22	July 11
Fiddle Dock (*Rumex pulcher*), fl.	July 4	3	June 25	July 10
Hemlock (*Conium maculatum*), fl.	July 4	4	June 17	July 22
Rye ripe and ready for cutting	July 4			
Sulphur-coloured Trefoil (*Trifolium ochroleucum*), fl. ...	July 4	2	June 26	July 13
White Lily (*Lilium candidum*), fl.	July 4	10	June 14	July 21
Allecula sulphurea, ap.	July 5			
Chaffinch (*Fringilla cælebs*), sg. ceas.	July 5	11	June 27	July 17
Hen-harrier (*Circus cyaneus*), hatches	July 5			
Scarlet-tiger Moth (*Hypercompa dominula*), ap. ...	July 5			
Silver-studded-blue Butterfly (*Polyommatus argus*), ap.	July 5			
Tree Pipit (*Anthus arboreus*), lays a second time ...	July 5			
Burnet-saxifrage (*Pimpinella saxifraga*), fl.	July 5	2	June 21	July 20
Cultivated Oat (*Avena sativa*), fl.	July 5	4	June 13	July 24
Flowering-rush (*Butomus umbellatus*), fl.	July 5	6	June 15	July 19
Money-wort (*Lysimachia nummularia*), fl.	July 5	5	June 20	July 17

([a]) They begin to return about this period; but the numbers keep gradually increasing as the summer advances.

JULY.

PHENOMENA.	Date of occurrence Mean	Earliest	Latest
Sharp Dock (*Rumex conglomeratus*), fl.	July 5 2		
Yellow Bedstraw (*Galium verum*), fl.	July 5 9	June 22	July 18
Young broods of Partridges fledged	July 6		
Yellow-underwing Moth (*Triphæna pronuba*), ap. ...	July 6 4	June 24	July 18
Young broods of Yellow-hammers fledged	July 6 2	June 17	July 26
Creeping Thistle (*Carduus arvensis*), fl.	July 6 7	June 22	July 17
Dwarf Thistle (*Carduus acaulis*), fl.	July 6 7	June 21	July 17
Hairy Brome-grass (*Bromus asper*), fl.	July 6		
Glow-worm (*Lampyris noctiluca*), shines	July 7 2	June 25	July 19
Lesser Whitethroat (*Curruca sylviella*), sg. ceas. ...	July 7 9	June 26	July 30
Great Water-plantain (*Alisma plantago*), fl.	July 7 6	June 12	July 26
Heath false Brome-grass (*Brachypodium pinnatum*), fl.	July 7 3	June 29	July 16
Marsh Helleborine (*Epipactis palustris*), fl.	July 7		
Lappet Moth (*Gastropacha quercifolia*), ap.	July 8		
Young broods of the common Lizard (*Zootoca vivipara*), ap.	July 8		
Ringlet Butterfly (*Hipparchia hyperanthus*), ap. ...	July 8 7	June 17	July 20
Everlasting-pea (*Lathyrus latifolius*), fl.	July 8 4	June 13	July 23
Great Bindweed (*Convolvulus sepium*), fl.	July 8 9	June 28	July 20
Marjoram (*Origanum vulgare*), fl.	July 8		

Reed Meadow-grass (*Glyceria aquatica*), fl.	July 8			
Slender false Brome-grass (*Brachypodium sylvaticum*), fl.	July 8	2	July 6	July 11
Spiked Speedwell (*Veronica spicata*), fl.	July 8	3	June 26	July 18
Trees make their Midsummer shoots ([a])	July 8	6	June 29	July 27
White Jasmine (*Jasminum officinale*), fl.	July 8	6	June 29	July 15
Shore Beetle (*Necrodes littoralis*), ap.	July 9	2	July 1	July 17
Cat-mint (*Nepeta cataria*), fl.	July 9	14	June 24	Aug. 2
Hawkweed Picris (*Picris hieracoides*), fl.	July 9	5	July 5	July 19
Song Thrush (*Turdus musicus*), lays a second time ...	July 10			
Young broods of spotted Flycatchers fledged	July 10			
Tree Pipit (*Anthus arboreus*), sg. ceas.	July 10	10	June 20	July 21
Branched Bur-reed (*Sparganium ramosum*), fl. ...	July 10	5	June 27	Aug. 3
Clove Pink (*Dianthus caryophyllus*), fl.	July 10	8	June 18	Aug. 1
Water-soldier (*Stratiotes aloides*), fl.	July 10			
Wild Succory (*Cichorium intybus*), fl.	July 10	9	June 15	July 25
Young Kestrels fledged ([b])	July 11			
Large-eggar Moth (*Lasiocampa quercus*), ap.	July 11	2	June 20	Aug. 2
Autumnal Oporinia (*Leontodon autumnalis*), fl. ...	July 11	6	June 25	Aug. 1
Great Mullein (*Verbascum thapsus*), fl.	July 11	6	June 26	Aug. 12

([a]) This is only a general observation, and perhaps of not much value. In repeating it, it would be well to mention in each case the particular tree.

([b]) Probably second broods, as it will be seen by a former part of this calendar that the Kestrel lays April 22. I do not find mention, however, in authors, of this species breeding twice in the season.

JULY.

Phenomena.	Date of occurrence: Mean		Earliest	Latest
Purple Loosestrife (*Lythrum salicaria*), fl.	July 11	12	June 20	Aug. 2
Spotted Persicaria (*Polygonum persicaria*), fl. ...	July 11	3	June 18	Aug. 1
Garden Warbler (*Curruca hortensis*), sg. ceas.... ...	July 12	8	June 28	July 20
Magpie Moth (*Abraxas grossulariata*), ap.	July 12	6	June 29	July 22
Silver-washed-fritillary Butterfly (*Argynnis paphia*), ap.	July 12			
Buckwheat (*Fagopyrum esculentum*), fl.	July 12	4	June 17	Aug. 1
Great Water-dock (*Rumex hydrolapathum*), fl. ...	July 12	3	July 4	July 19
Houseleek (*Sempervivum tectorum*), fl.	July 12			
Marsh Ragwort (*Senecio aquaticus*), fl.	July 12	3	July 4	July 21
Nettle-leaved Bell-flower (*Campanula trachelium*), fl.	July 12	3	June 21	July 31
Small-flowered Willow-herb (*Epilobium parviflorum*), fl.	July 12	7	June 28	July 27
Vervain (*Verbena officinalis*), fl.	July 12	10	June 17	Aug. 4
Wild Basil (*Calamintha clinopodium*), fl.	July 12	7	July 7	July 22
Hoplia argentea, ap.	July 13	3	June 24	July 31
Fine-leaved Water-dropwort (*Œnanthe phellandrium*), fl.	July 13	2	July 10	July 16
Common Centaury (*Erythræa pulchella*), fl.	July 14	3	July 10	July 20
Drinker Moth (*Odonestis potatoria*), ap.	July 14	4	June 30	July 26
Great Willow-herb (*Epilobium hirsutum*), fl.	July 14	10	June 28	July 30

Traveller's-joy (*Clematis vitalba*), fl.	July 14	3	July 8	July 19
Black Nightshade (*Solanum nigrum*), fl.	July 15	4	June 17	Aug. 1
Gooseberries ripe	July 15	6	June 28	July 25
Peach-leaved Bell-flower (*Campanula persicifolia*), fl.	July 15			
Upright Hedge-parsley (*Torilis anthriscus*), fl. ...	July 15	4	July 1	July 31
White Goose-foot (*Chenopodium album*), fl.	July 15	4	July 1	Aug. 7
White Horehound (*Marrubium vulgare*), fl.	July 15	5	June 7	Aug. 18
White Poppy (*Papaver somniferum*), fl.	July 15	5	June 30	July 30
Blackbird (*Turdus merula*), sg. ceas.	July 16	11	June 29	July 28
Dark-green-fritillary Butterfly (*Argynnis aglaia*), ap.	July 16			
Large-heath Butterfly (*Hipparchia tithonus*), ap. ...	July 16	8	July 2	Aug. 1
V Moth (*Grammatophora vauaria*), ap.	July 16	2	July 2	July 30
Broad-leaved Water-parsnep (*Sium latifolium*), fl. ...	July 16	3	July 6	July 25
Large-flowered St John's-wort (*Hypericum calycinum*), fl.	July 16	6	July 3	July 28
Marsh Woundwort (*Stachys palustris*), fl.	July 16	7	July 5	Aug. 1
Procumbent Water-parsnep (*Heliosciadium nodiflorum*), fl.	July 16	2	July 6	July 25
Spear Thistle (*Carduus lanceolatus*), fl.	July 16	6	July 2	Aug. 2
Burnished-brass Moth (*Plusia chrysitis*), ap.	July 17			
Common Hemp-nettle (*Galeopsis tetrahit*), fl. ...	July 17	9	July 8	July 30
Eyebright (*Euphrasia officinalis*), fl.	July 17	5	June 18	Aug. 12
Narrow-leaved Water-parsnep (*Sium angustifolium*), fl.	July 17	4	July 1	Aug. 9

JULY.

Phenomena.	Date of occurrence: Mean		Earliest	Latest
Pale-flowered Persicaria (*Polygonum lapathifolium*), fl.	July 17	3	July 10	Aug. 1
Simple Bur-reed (*Sparganium simplex*), fl.	July 17	4	June 27	Aug. 10
Square-stalked St John's-wort (*Hypericum quadrangulum*), fl.	July 17	8	July 2	Aug. 4
Cotton-thistle (*Onopordon acanthium*), fl.	July 18	5	July 1	Aug. 12
Dark-arches Moth (*Xylophasia polyodon*), ap. ...	July 18			
Spreading Hedge-parsley (*Torilis infesta*), fl.	July 18			
Tutsan (*Hypericum androsæmum*), fl.	July 18	8	June 30	Aug. 4
Wren (*Troglodytes europœus*), second broods fledged...	July 18			
Barred-lackey Moth (*Clisiocampa neustria*), ap. ...	July 19	3	July 10	July 23
Chalk-hill-blue Butterfly (*Polyommatus corydon*), ap.	July 19	8	July 5	Aug. 12
Reed Bunting lays a second time	July 19			
Song Thrush (*Turdus musicus*), sg. ceas.	July 19	9	June 26	July 27
Whitethroat (*Curruca cinerea*), sg. ceas.	July 19	9	June 28	July 28
Blue Funkia (*Funkia ovata*), fl.	July 19	5	July 6	July 25
Great Reed-mace (*Typha latifolia*), fl. ([a])	July 19	4	July 8	Aug. 1
Small Teasel (*Dipsacus pilosus*), fl.	July 19	10	June 27	Aug. 7
Garden-tiger Moth (*Arctia caja*), ap.	July 20	3	July 15	July 31
Goat Moth (*Cossus ligniperda*), ap.	July 20	2	July 15	July 25
Amphibious Persicaria (*Polygonum amphibium*), fl. ...	July 20	4	June 28	Aug. 3

Corn Feverfew (*Matricaria inodora*), fl.	July 20	5	July 12	July 27
Fool's-parsley (*Æthusa cynapium*), fl.	July 20	2	July 17	July 24
Hemp-agrimony (*Eupatorium cannabinum*), fl. ...	July 20	10	July 4	Aug. 14
Red Eyebright (*Bartsia odontites*), fl.	July 20	5	July 10	July 28
Arrowhead (*Sagittaria sagittifolia*), fl.	July 21	8	June 28	Aug. 10
Musk Beetle (*Cerambyx moschatus*), ap.	July 21			
Tit Lark (*Anthus pratensis*), sg. ceas.	July 21	4	July 16	July 27
Willow Warbler (*Sylvia trochilus*), sg. ceas.	July 21	10	June 26	Sept. 13
Humming-bird Hawkmoth (*Macroglossa stellatarum*), ap.	July 22	7	June 29	Aug. 6
Necrophorus vestigator, ap.	July 22			
Sedge Warbler (*Salicaria phragmitis*), sg. ceas. ...	July 22	10	June 28	Aug. 30
Burdock (*Arctium lappa*), fl.	July 22	11	June 29	Aug. 13
Great-yellow Loosestrife (*Lysimachia vulgaris*), fl. ...	July 22	5	July 11	Aug. 5
Swallow-tail Moth (*Ourapteryx sambucaria*), ap. ...	July 24	2	July 17	July 31
Fennel (*Fœniculum officinale*), fl.	July 24	5	July 5	Aug. 11
Flea-bane (*Pulicaria dysenterica*), fl.	July 24	15	July 2	Aug. 8
Second broods of Goldfinches fledged	July 25			
Large-marsh Grasshopper (*Locusta flavipes*), ap. ...	July 25			
Apricots ripe	July 25	4	July 17	Aug. 4
Common Star-thistle (*Centaurea calcitrapa*), fl. ...	July 25	5	July 2	Aug. 19
Red Hemp-nettle (*Galeopsis ladanum*), fl.	July 25	6	July 15	Aug. 8

(a) [Usually known as Bullrush.]

JULY.

Phenomena.	Date of occurrence: Mean		Earliest	Latest
Grayling Butterfly (*Hipparchia semele*), ap.	July 26	3	July 12	Aug. 7
Great-green Acrida (*Acrida viridissima*), stridulous note heard	July 26	6	July 12	Aug. 9
Hedge Sparrow (*Accentor modularis*), sg. ceas. ...	July 26	9	July 14	Aug. 23
Dwarf Elder (*Sambucus ebulus*), fl.	July 26	2	July 19	Aug. 2
Blackcap (*Curruca atricapilla*), sg. ceas.	July 27	11	June 29	Aug. 18
Turtle-dove (*Columba turtur*), last heard	July 27	7	July 18	Aug. 16
Bastard Stone-parsley (*Sison amomum*), fl.	July 27	7	July 15	Aug. 9
Common Grasshopper (*Locusta biguttula*), crinks ...	July 28	4	July 19	Aug. 9
Small-skipper Butterfly (*Pamphila linea*), ap. ...	July 28	3	July 18	Aug. 11
Young broods of Swifts (*Cypselus apus*), fledged ...	July 28			
Common Feverfew (*Chrysanthemum parthenium*), fl. ...	July 28	3	July 20	Aug. 6
Gipsy-wort (*Lycopus europæus*), fl.	July 28	6	July 21	Aug. 5
Hairy Mint (*Mentha sativa*), fl.	July 28	9	July 11	Aug. 14
Knotted Spurrey (*Sagina nodosa*), fl.	July 28	2	July 27	July 29
Wild Teasel (*Dipsacus sylvestris*), fl.	July 28	11	July 8	Aug. 13
Common Calamint (*Calamintha officinalis*), fl. ...	July 29	3	July 12	Aug. 17
Wheat cut	July 30	16	July 16	Aug. 18
Wild Angelica (*Angelica sylvestris*), fl.	July 30	5	July 25	Aug. 11
Admiral Butterfly (*Vanessa atalanta*), ap.	July 31	6	June 30	Aug. 14

AUGUST.

Second broods of Swallows fledged	Aug. 1			
Hoary Ragwort (*Senecio erucifolius*), fl.	Aug. 1	7	July 19	Aug. 15
Wormwood (*Artemisia absinthium*), fl.	Aug. 1	3	July 24	Aug. 11
Large-flowered Hemp-nettle (*Galeopsis versicolor*), fl.	Aug. 4	2	Aug. 2	Aug. 6
Mugwort (*Artemisia vulgaris*), fl.	Aug. 4	4	July 26	Aug. 8
Mushrooms (*Agaricus campestris*), abound	Aug. 5	3	June 20	Sept. 20
Globe Thistle [*Echinops* ?], fl.	Aug. 6			
Honeysuckle berries ripe	Aug. 6	4	July 29	Aug. 13
Bracts of the Lime fall ([a])	Aug. 7	6	July 19	Sept. 8
Coal Titmouse (*Parus ater*), note ceas.	Aug. 8	3	July 13	Sept. 22
Hornet-fly (*Asilus crabroniformis*), ap.	Aug. 8			
Swift (*Cypselus apus*), last seen	Aug. 8	8	July 29	Aug. 23
Carline-thistle (*Carlina vulgaris*), fl.	Aug. 8	3	Aug. 6	Aug. 10
Common Linnet (*Linota cannabina*), sg. ceas. ...	Aug. 9	8	July 18	Aug. 30
Purple Melic-grass (*Molinia cærulea*), fl.	Aug. 9	2	July 27	Aug. 22
Red Goose-foot (*Chenopodium rubrum*), fl.	Aug. 9			
Silver-spotted-skipper Butterfly (*Pamphila comma*), ap.	Aug. 10			
Ringdove lays a second time ([b])	Aug. 12			
Yellow-hammer (*Emberiza citrinella*), sg. ceas. ...	Aug. 12	9	July 29	Aug. 27

([a]) This is synchronous with the ripening of the fruit, to which the bracts are attached.
([b]) Perhaps the third, instead of the second time.

AUGUST.

Phenomena.	Date of occurrence: Mean		Earliest	Latest
Soapwort (*Saponaria officinalis*), fl.	Aug. 12	3	July 25	Sept. 1
Second broods of house Martins fledged	Aug. 13			
Zabrus gibbus, ap.	Aug. 13	4	July 21	Sept. 11
Orpine (*Sedum telephium*), fl.	Aug. 13	2	Aug. 12	Aug. 15
Goldfinch (*Carduelis elegans*), sg. ceas.	Aug. 14	5	Aug. 3	Aug. 31
Snowberries ripe	Aug. 14	2	Aug. 12	Aug. 17
Large-black Staphyline (*Goërius olens*), ap.	Aug. 15	5	July 6	Sept. 15
Swallows and Martins begin to congregate	Aug. 15	12	July 23	Sept. 9
Barley cut	Aug. 15	5	Aug. 1	Aug. 30
Artichoke (*Cynara scolymus*), fl.	Aug. 16			
Greenfinch (*Coccothraustes chloris*), sg. ceas.	Aug. 16	9	July 24	Sept. 8
Jargonelle Pears ripe	Aug. 16			
Red Bryony berries ripe	Aug. 16	5	July 28	Sept. 21
Devil's-bit Scabious (*Scabiosa succisa*), fl.	Aug. 18	5	Aug. 5	Aug. 31
Common Tansy (*Tanacetum vulgare*), fl.	Aug. 19			
Redbreast (*Erithaca rubecula*), sg. reass.	Aug. 20	8	July 29	Sept. 10
Small-copper Butterfly (*Lycæna phlæas*), ap.	Aug. 20	2	Aug. 17	Aug. 23
Woolly-headed Thistle (*Carduus eriophorus* , fl.	Aug. 21	4	Aug. 12	Sept. 2
Gold-spot Moth (*Plusia festucæ*), ap.	Aug. 23			
Starlings collect in flocks	Aug. 23	12	July 12	Oct. 9

Winged Ants migrate	Aug. 24	12	July 26	Sept. 28
Peaches ripe	Aug. 25	4	Aug. 14	Sept. 7
Martins collect in great numbers on the roofs of houses	Aug. 29			
Yew-berries ripe	Aug. 30	5	July 29	Sept. 25
Autumnal Gentian (*Gentiana amarella*), fl.	Aug. 31	4	Aug. 12	Sept. 8

SEPTEMBER.

Clouded-yellow Butterfly (*Colias edusa*), ap.	Sept. 1	2	Aug. 23	Sept. 11
Chaffinch (*Fringilla cœlebs*), sg. reass.	Sept. 2	5	Aug. 22	Sept. 19
Barberries ripe	Sept. 2	2	Aug. 24	Sept. 11
Meadow-saffron (*Colchicum autumnale*), fl.	Sept. 3			
Stock-dove (*Columba œnas*), note ceas.	Sept. 5			
Cuckow-pint (*Arum maculatum*), berries ripe ...	Sept. 5	4	Aug. 24	Sept. 17
Gossamer floats	Sept. 6	9	Aug. 18	Sept. 23
Red-underwing Moth (*Catocala nupta*), ap.	Sept. 6	6	Aug. 11	Oct. 1
Swallow (*Hirundo rustica*), sg. ceas.	Sept. 7	7	Aug. 11	Oct. 2
Elder-berries ripe	Sept. 7	6	Aug. 11	Oct. 1
Hawthorn berries ripe	Sept. 8	5	Sept. 1	Sept. 20
Dog-rose casts its leaves	Sept. 9			
Great Titmouse (*Parus major*), sg. reass.	Sept. 10	3	Sept. 6	Sept. 17
House-flies swarm in windows	Sept. 12	3	Sept. 5	Sept. 17
Lime turns yellow	Sept. 14	9	Aug. 17	Sept. 29

SEPTEMBER.

Phenomena.	Date of occurrence Mean		Earliest	Latest
House Sparrows collect in large flocks	Sept. 15			
Vapourer Moth (*Orgyia antiqua*), ap.	Sept. 15	2	Sept. 8	Sept. 23
Gold-crested Wren (*Regulus cristatus*), sg. reass. ...	Sept. 16	6	Sept. 9	Oct. 1
Drone-fly (*Eristalis tenax*), enters houses	Sept. 18	12	Aug. 22	Oct. 15
Peewits collect in flocks	Sept. 18	3	Aug. 25	Oct. 5
Dotterel makes its autumnal passage	Sept. 19	3	Sept. 17	Sept. 22
Acorns fall	Sept. 19			
Chiffchaff (*Sylvia hippolais*), note ceas.	Sept. 20	5	Sept. 9	Sept. 29
Tutsan (*Hypericum androsæmum*), turns brown ...	Sept. 21			
Goldfinches collect in flocks	Sept. 22	7	Sept. 5	Oct. 29
Laurestine (*Viburnum tinus*), fl.	Sept. 22	8	Aug. 18	Dec. 16
Herald Moth (*Calyptra libatrix*), ap.	Sept. 23	5	Aug. 26	Nov. 9
Beech-mast falls	Sept. 24	6	Sept. 4	Oct. 22
Horse-chestnut turns brown	Sept. 24	4	Sept. 12	Oct. 3
Lime leaves begin to fall	Sept. 24	9	Sept. 2	Oct. 7
Wych Elm leaves begin to fall	Sept. 24	4	Sept. 7	Oct. 20
Syringa (*Philadelphus coronarius*), turns yellow ...	Sept. 27	7	Sept. 2	Oct. 29
Acrida varia, ap.	Sept. 28	4	Sept. 14	Oct. 30
Ringdove (*Columba palumbus*), note ceas.	Sept. 28	7	Sept. 21	Oct. 9
Horse-chestnuts fall	Sept. 28	4	Sept. 16	Oct. 8

Autumn-green-carpet Moth (*Harpalyce miata*), ap. ...	Sept. 29	2	Sept. 25	Oct. 3
Hedge Sparrow (*Accentor modularis*), sg. reass. ...	Sept. 29	6	Sept. 14	Oct. 12
Ivy (*Hedera helix*), fl.	Sept. 29	13	Sept. 5	Nov. 9
Walnuts ripe	Sept. 30	2	Sept. 25	Oct. 5

OCTOBER.

Common Snipe (*Scolopax gallinago*), ap. in plenty ([a])	Oct. 1	7	Aug. 20	Nov. 14
Beech turns yellow	Oct. 1	10	Sept. 6	Oct. 23
Birch turns yellow	Oct. 1	5	Sept. 10	Oct. 16
Sycamore leaves fall	Oct. 2	4	Sept. 15	Oct. 17
Jack-Snipe (*Scolopax gallinula*), arrives	Oct. 3	6	Sept. 20	Oct. 28
Martins (*Hirundo urbica*), bulk departed	Oct. 4	4	Sept. 29	Oct. 10
Horse-chestnut leaves begin to fall	Oct. 4	6	Aug. 27	Oct. 17
Sloes ripe	Oct. 4	3	Sept. 5	Nov. 8
Virginian-Creeper (*Ampelopsis hederacea*), turns red ...	Oct. 4	8	Sept. 13	Oct. 28
Buntings (*Emberiza miliaria*), collect in flocks ...	Oct. 6			
Linnets (*Linota cannabina*), collect in flocks	Oct. 7	9	Sept. 6	Nov. 13
Maple turns yellow	Oct. 7	2	Oct. 1	Oct. 13
Trees in general assume their autumnal tints ...	Oct. 7	12	Sept. 22	Oct. 22
Birch leaves fall	Oct. 8	4	Sept. 30	Oct. 19
Cherry leaves fall	Oct. 8			

([a]) Many Snipe remain with us the whole summer, but there is an accession of numbers in the autumn.

OCTOBER.

Phenomena.	Date of occurrence			
	Mean		Earliest	Latest
Wheat sown	Oct. 9	9	Sept. 23	Oct. 23
White Poplar leaves begin to fall	Oct. 10	5	Sept. 25	Oct. 19
Crab-apples ripe and falling	Oct. 12			
Elm turns yellow	Oct. 12	5	Oct. 4	Oct. 20
Hazel turns yellow	Oct. 12	4	Oct. 5	Oct. 20
Maple leaves fall	Oct. 12	2	Oct. 7	Oct. 17
Walnut leaves begin to fall	Oct. 12	2	Oct. 2	Oct. 23
Aspen leaves fall	Oct. 13	2	Oct. 6	Oct. 20
Beech leaves begin to fall	Oct. 13	3	Oct. 7	Oct. 20
Elder leaves fall	Oct. 13			
Ladybird (*Coccinella septempunctata*), hybernates ...	Oct. 14	2	Oct. 8	Oct. 20
Swallow (*Hirundo rustica*), last seen	Oct. 14	15	Sept. 28	Oct. 31
Ash leaves begin to fall ([a])	Oct. 14	6	Sept. 20	Oct. 28
Martin (*Hirundo urbica*), last seen	Oct. 17	13	Oct. 2	Nov. 4
Dogwood turns red	Oct. 17			
Lombardy Poplar leaves fall	Oct. 18	4	Oct. 11	Nov. 3
Virginian-Creeper leaves fall	Oct. 18	5	Oct. 13	Oct. 19
Yellow-hammer (*Emberiza citrinella*), sg. reass. ...	Oct. 19	3	Oct. 14	
Lime stript of its leaves	Oct. 20	10	Sept. 22	Oct. 29
Gull comes inland	Oct. 22			

Honeysuckle leaves fall	Oct. 22	2	Oct. 12	Nov. 1
Golden Plover (*Charadrius pluvialis*), arrives ...	Oct. 24	2	Sept. 22	Nov. 25
Short-eared Owl (*Otus brachyotos*), arrives	Oct. 25	3	Sept. 21	Nov. 15
Hazel leaves begin to fall	Oct. 26	2	Oct. 17	Nov. 4
Flocks of Wild-Geese (*Anser segetum*), arrive	Oct. 28	12	Oct. 7	Nov. 29
Woodcock (*Scolopax rusticola*), arrives	Oct. 28	12	Oct. 14	Nov. 18
Elm leaves begin to fall	Oct. 28	6	Oct. 18	Nov. 15
Virginian-Creeper stript	Oct. 28	5	Oct. 24	Nov. 1
Walnut stript of its leaves	Oct. 28	6	Oct. 5	Nov. 10
Wild-Duck (*Anas boschas*), arrives	Oct. 29	3	Oct. 7	Nov. 21
Horse-chestnut stript	Oct. 30	7	Oct. 19	Nov. 15
Whitethorn leaves fall	Oct. 30	2	Oct. 26	Nov. 3
Skylark (*Alauda arvensis*), sg. ceas.	Oct. 31	5	Oct. 9	Dec. 10
Sycamore stript of leaves	Oct. 31	4	Oct. 21	Nov. 6

NOVEMBER.

Missel Thrush (*Turdus viscivorus*), sg. reass.	Nov. 2	5	Sept. 20	Dec. 11
Ash stript	Nov. 3	6	Oct. 27	Nov. 11
Maple stript	Nov. 3	2	Nov. 1	Nov. 5
Plane leaves fall	Nov. 3			

(a) Linnæus observes of the Ash, "*Fraxinus inter primas defoliatur et inter ultimas frondescit.*" (Phil. Bot.) This is often the case. The defoliation, however, depends a good deal upon the time of occurrence of the first sharp frost, which frequently brings all the leaves down together, so that a tree is nearly stript in a night. The consequence is, that the leaves fall *green*, unlike those of any other tree.

NOVEMBER.

Phenomena.	Date of occurrence: Mean		Earliest	Latest
White Poplar stript	Nov. 3	4	Oct. 30	Nov. 7
Cherry-tree stript	Nov. 5			
Crab-apple stript	Nov. 5			
Guelder-rose stript	Nov. 5	2	Nov. 1	Nov. 9
Laburnum stript	Nov. 6	3	Nov. 4	Nov. 10
Red Currant stript	Nov. 6	4	Oct. 29	Nov. 13
Syringa stript	Nov. 6	4	Oct. 29	Nov. 13
Whitethorn stript	Nov. 6	3	Nov. 4	Nov. 9
Hooded Crow (*Corvus cornix*), arrives	Nov. 7	3	Oct. 9	Dec. 5
Larch turns yellow	Nov. 8			
Apple-tree stript	Nov. 9	4	Nov. 4	Nov. 18
Bladdernut stript	Nov. 9			
Hornbeam stript	Nov. 9	3	Nov. 4	Nov. 16
Lilac stript	Nov. 9	4	Nov. 3	Nov. 20
Beech stript	Nov. 10	6	Oct. 28	Nov. 16
Bunting (*Emberiza miliaria*), note ceas.	Nov. 10	2	Oct. 11	Dec. 9
Birch stript	Nov. 12	3	Nov. 8	Nov. 16
Hazel stript	Nov. 13	3	Nov. 1	Nov. 20
Woodpigeons collect in flocks	Nov. 13	8	Oct. 11	Dec. 20
Oak stript	Nov. 14	4	Nov. 4	Nov. 26

Gooseberry stript	Nov. 15	3	Nov. 4	Nov. 21
Larch leaves fall	Nov. 15			
Lombardy Poplar stript	Nov. 15	2	Nov. 8	Nov. 22
Teal (*Anas crecca*), arrives	Nov. 16	2	Oct. 3	Dec. 31
Titmice draw near houses	Nov. 16	8	Oct. 17	Dec. 17
Black currant stript	Nov. 16	2	Nov. 11	Nov. 21
Vine stript	Nov. 16			
Fieldfare (*Turdus pilaris*), arrives	Nov. 17	12	Oct. 26	Dec. 21
Apricot stript	Nov. 17	3	Nov. 15	Nov. 19
Wych Elm stript	Nov. 18			
Fig stript	Nov. 19			
Peach stript	Nov. 19	2	Nov. 13	Nov. 26
Grey Wagtail (*Motacilla boarula*), arrives	Nov. 21	4	Oct. 10	Jan. 8
Redwing (*T. iliacus*), arrives	Nov. 21	3	Oct. 29	Dec. 21
Elder stript	Nov. 21			
Song Thrush (*Turdus musicus*), sg. reass.	Nov. 24	8	Nov. 5	Dec. 13
Elm stript	Nov. 24	6	Nov. 22	Nov. 30
Larch stript	Nov. 24			
Jargonelle Pear stript	Nov. 26			
Trees everywhere stript of leaves	Nov. 29	8	Nov. 20	Dec. 12

DECEMBER.

PHENOMENA.	Date of occurrence: Mean		Earliest	Latest
Pipistrelle Bat last seen abroad	Dec. 2	8	Oct. 25	Dec. 31
Skylarks collect in flocks	Dec. 7	9	Nov. 26	Dec. 13
Christmas Rose (*Helleborus niger*), fl.	Dec. 8			
Greenfinches collect in flocks	Dec. 15	2	Dec. 13	Dec. 17
Chaffinches collect in flocks	Dec. 25			
Marsh Titmouse (*Parus palustris*), sg. reass.	Dec. 25			

ALPHABETICAL ARRANGEMENT

OF THE

PERIODIC PHENOMENA IN THE FOREGOING CALENDAR,

WITH A REFERENCE TO THE MEAN DATE OF OCCURRENCE.

The time of flowering (fl.) *is determined by the visibility of the anthers.*
The time of leafing (l.) *by the exposure of the upper surface of the leaves.*
For the meaning of other abbreviations see page 2.

A

Abraxas grossulariata, ap. ... ... July 12
Accentor modularis, see Hedge-Sparrow
Acer campestre, see Maple
,, *pseudoplatanus*, see Sycamore
Achillea millefolium, fl. June 29
Achyrophorus maculatus, fl. ... ... June 26
Aconite, winter, flowers Jan. 26
Aconitum napellus, fl. May 26
Acorns, fall ... ... Sept. 19
Acrida, great green, stridulous note heard ... July 26
Acrida varia, ap. ... Sept. 28
,, *viridissima*, heard July 26
Acronycta psi, ap. ... June 16
Ægopodium podagraria, fl. ... ... June 12
Æsculus hippocastanum, see Horse-Chestnut
Æshna grandis, ap. ... June 10
Æthusa cynapium, fl. July 20
Agaricus campestris, abounds ... ... Aug. 5
Agrimonia eupatoria, fl. July 1
Agrimony, common, fl. July 1
Ajuga reptans, fl. ... May 3
Alauda arvensis, see Skylark
Alder, fl. ... ... Feb. 28
,, leafs ... ... Apr. 11
Alisma plantago, fl. ... July 7
Alkanet, evergreen, fl. Apr. 14
Allecula sulphurea, ap. July 5
Alliaria officinalis, fl.... Apr. 22
Almond, fl.... ... ... Mar. 28
Alnus glutinosa, see Alder
Alopecurus pratensis, fl. May 14
,, *agrestis*, fl. May 24
Alucita hexadactyla, ap. Apr. 5
American-Cowslip, fl. Apr. 26
Ampelopsis, see Virginian-creeper
Amygdalus communis, fl. Mar. 28
,, *persica*, see Peach
Anagallis arvensis, fl. June 8
Anas boschas, see Duck and Wild duck
Anas crecca, arrives ... Nov. 16

B

D

F

I

M

N

O

P

Pheasant, utters its spring crow ... Apr. 8
,, lays ... ... May 6
Pheasants, young, hatched ... ... June 4
Philadelphus coronarius, see Syringa
Philomela luscinia, see Nightingale
Phœnicura ruticilla, see Redstart
Pica caudata, builds ... Mar. 22
Picris, hawkweed, fl. July 9
Picris hieracoides, fl. July 9
Picus viridis, cries ... Mar. 25
Pigeon, house, lays ... Feb. 11
,, ,, hatches Mar. 8
Pilewort, fl. ... ... Feb. 28
Pimpernel, scarlet, fl. June 8
Pimpinella saxifraga, fl. July 5
Pinguicula vulgaris, fl. May 25
Pink, clove, fl. ... ... July 10
,, common garden, fl. May 30
Pipit, tree, first heard Apr. 21
,, lays ... ... May 18
,, lays a second time July 5
,, song ceases ... July 10
Plane, l. ... ... May 8
,, leaves fall ... Nov. 3
Plantago lanceolata, fl. Apr. 27
,, *major*, fl. ... ... June 28
,, *media*, fl. ... ... May 27
Plantain, great, fl. ... June 28
,, hoary, fl. ... May 27
,, ribwort, fl. ... Apr. 27
Platanus orientalis, l. May 8
Plecotus auritus, comes abroad ... ... May 7
Plover, golden, arrives Oct. 24
,, great, first heard or seen ... ... Apr. 13
Plum, l. ... ... Apr. 5
,, fl. ... ... Apr. 7

Plusia chrysitis, ap. ... July 17
,, *festucæ*, ap. ... Aug. 23
,, *gamma*, ap. ... June 19
Poa compressa, fl. ... May 31
,, *pratensis*, fl. ... May 28
,, *trivialis*, fl. ... ... June 12
Pœcilus cupreus, ap. ... Apr. 19
Polemonium cæruleum, fl. ... ... June 12
Polygala vulgaris, fl. May 15
Polygonatum multiflorum, fl. ... ... May 15
Polygonum amphibium, fl. ... ... July 20
,, *bistorta*, fl. ... May 25
,, *lapathifolium*, fl. July 17
,, *persicaria*, fl. ... July 11
Polyommatus agrestis, ap. ... ... May 26
,, *alexis*, ap. ... June 4
,, *alsus*, ap. ... ... June 12
,, *argus*, ap. ... ... July 5
,, *corydon*, ap. ... July 19
Pontia brassicæ, ap. ... Apr. 19
,, *cardamines*, ap. May 11
,, *napi*, ap. ... ... Apr. 25
,, *rapæ*, ap. ... ... Apr. 17
Poplar, black, fl. ... Apr. 4
,, leafs ... ... Apr. 23
Poplar, Lombardy, fl. Apr. 4
,, leafs ... ... Apr. 23
,, leaves fall ... Oct. 18
,, stript ... ... Nov. 15
Poplar, white, fl. ... Mar. 24
,, leafs ... ... Apr. 18
,, leaves fall ... Oct. 10
,, stript ... ... Nov. 3
Poppy, common, red, fl. ... ... June 4
,, large oriental, fl. June 6
,, long-prickly-headed, fl. ... June 3
,, white, fl. ... ... July 15

Q

R

S

U

V

W

X

Y

Z

Printed in the United States
By Bookmasters